W9-AWJ-671

The Invisible Universe

George B. Field, formerly Director of the
Harvard-Smithsonian Center for Astrophysics,
is now the Robert Wheeler Willson Professor
of Applied Astronomy at Harvard University
and Senior Physicist at the Smithsonian
Astrophysical Observatory.

Eric J. Chaisson, formerly a member
of the Harvard faculty, is now
Professor of Astronomy at Haverford College.

Thomas P. Stephenson, formerly Director of the
Harvard-Smithsonian Center for Astrophysics
Image Processing Facility, is currently Manager of
The Analytic Sciences Corporation (TASC)
Image Processing Laboratory.

The Invisible Universe

Probing the frontiers of astrophysics

by

GEORGE B. FIELD and ERIC J. CHAISSON

Astronomical Images and Descriptions

by

Thomas P. Stephenson

Birkhäuser
Boston · Basel · Stuttgart

Library of Congress Cataloging in Publication Data

Field, George B., 1929–
The invisible universe.

Includes index.
1. Astronomy. 2. Radio astronomy. 3. Astrophysics.
I. Chaisson, Eric. II. Title.
QB43.2.F54 1984 520 84-24621
ISBN 0-8176-3235-2

ISBN 0-8176-3235-2
ISBN 3-7643-3235-2

Printed in the USA

A B C D E F G H

To Susan and Lola

Table of Contents

Preface

...And indeed, these latest centuries merit praise because it is during them that the arts and sciences, discovered by the ancients, have been reduced to so great and constantly increasing perfection through the investigations and experiments of clear-seeing minds. This development is particularly evident in the case of the mathematical sciences. Here, without mentioning various men who have achieved success, we must without hesitation and with the unanimous approval of scholars assign the first place to Galileo Galilei, Member of the Academy of the Lincei. This he deserves not only because he has effectively demonstrated fallacies in many of our current conclusions, as is amply shown by his published works, but also by means of the telescope (invented in this country but greatly perfected by him) he has discovered the four satellites of Jupiter, has shown us the true character of the Milky Way, and has made us acquainted with spots on the Sun, with the rough and cloudy portions of the lunar surface, with the threefold nature of Saturn, with the phases of Venus and with the physical character of comets. These matters were entirely unknown to the ancient astronomers and philosophers; so that we may truly say that he has restored to the world the science of astronomy and has presented it in a new light.

From the publisher's preface to Galileo's
Dialogues Concerning Two New Sciences, Leiden, 1638

Early in the Seventeenth Century, Galileo Galilei trained his newly invented telescope on the heavens, almost instantaneously creating a revolution in astronomy, as well as a breakthrough in human perception. Viewing for the first time blemishes on the Sun, rugged mountains on the Moon, and whole new worlds orbiting Jupiter, he demolished the Aristotelian philosophy of cosmic immutability. Indeed, the Universe changes; and so does our perception of it.

Nearly four centuries later, we are experiencing another period of unsurpassed scientific achievement – a revolution in which contemporary astronomers are revealing the invisible Universe as Galileo once revealed the visible Universe. In the last two decades, we have learned how to detect, measure, and analyze invisible radiation streaming to us from dark objects in space. And once again our perceptions are changing.

Consider these recent advances made with instruments that are sensitive to radiation invisible to the human eye:

- Perhaps as much as ninety percent of the matter in the Universe is dark, invisible even to large optical telescopes.
- The fiery creation of the Universe itself is revealed only by an invisible radio emission, called the cosmic background radiation.
- Sharply tuned radio emissions from atoms and molecules in space demonstrate a rich chemistry at work in the dark depths of space between the stars of our nighttime sky.
- Strong radio signals, emitted by the powerful cores of the quasi-stellar sources (or *quasars*), seem to imply that matter is being ejected at speeds greater than that of light.
- Infrared waves, emanating from warm, compressed clouds in interstellar space, indicate that new stars are forming at this moment at myriad sites within them.
- Ultraviolet radiation, sampled by telescopes orbiting above our hazy atmosphere, is helping us to understand the birth, maturity, and death of stars.
- X rays (as well as radio and infrared waves), emanating from the heart of our Milky Way Galaxy, can be explained by a colossal whirlpool of hot gases in orbit around a supermassive black hole.
- Gamma rays, radiating from compact regions near the center of our Galaxy, must originate by the annihilation of antimatter – the strange opposite of matter on Earth.

From Earth and planets to stars and galaxies, astronomers and astrophysicists have accelerated their exploration of previously uncharted space. Over the course of the past two decades we have begun to unravel the mystery of the greatest puzzle of all: the Universe itself.

We still ask the same questions as did the ancients about ourselves, our environment, and our place in the cosmic scheme of matter and life. We seek a deeper understanding of the starry points of light in the evening sky.

We strive to understand the origin and destiny of things. But our present attempts to find answers are aided by experimental tools made possible by modern technology: telescopes to gather information about the astronomical Universe; high-energy accelerators to penetrate the sub-microscopic world of elementary particles; automated space probes to gather data unavailable on Earth; and sophisticated computers to help keep pace with the increasing wealth of new information.

Computerized telescopes now operate, from the ground and from orbit, in each of the invisible domains of the electromagnetic spectrum. Radio telescopes as large as engineers can build them scan the Universe for signals that cannot be observed visually. State-of-the-art infrared detectors fly aboard high-altitude balloons, reconnaissance aircraft, and satellites, seeking the warm clouds that mark the birthplaces of stars in space. Telescopes aboard spacecraft map distant sources of potent ultraviolet, x-ray, and gamma-ray emission from previously unknown exotic astronomical objects. These are not merely passive probes like the pioneering satellites that marked the dawn of the Space Age a mere quarter-century ago, but complete orbiting observatories operated remotely by teams of scientists and engineers, much like major ground-based telescopes. Robot space probes also navigate through the system of known planets, telemetering to Earth visible and invisible images of those totally alien worlds.

We are in a period of grand technological progress, a time of learning, of groping in the darkness – more a time of exploration than of mature science. Sampling the Universe's rich spectrum of radiation to uncover its myriad forms of matter, we currently strive to fit our lode of recent discoveries into a newly emerging scientific philosophy. Galileo would have been pleased.

In this book, we aim to share with the interested layperson a timely survey of our new cosmos, much of which is inherently invisible. Discussing stars, galaxies, and especially the hidden matter among them, our work summarizes key discoveries made in this golden age of astronomy which we now share. In each chapter, we concentrate on what we have recently learned, including the main problems currently puzzling us; we then discuss how we might attack those problems with observational equipment and theoretical insight expected to emerge during the remaining years of this century.

Our book is inspired by the recently completed Report of the Astronomy Survey Committee, which recommends a program of astronomical research for the 1980s. The Report resulted from a two-year national study of the goals, aspirations, and objectives of the American astronomical community. Chaired by one of us (G.B.F.), the study was commissioned by the National Academy of Sciences at the request of the Federal Government.

We hope that this book, as well as the Report* from which it derives, will aid in developing long-term policies designed to unlock secrets of the Universe for the good of all humankind.

We are indebted to Drs. Robert Harlow, Alan Lightman, and Mark Stier for a careful reading of the manuscript, and to our illustrator, Tom Stephenson, for contributing to the section entitled "Astronomical Image Processing" that begins on page 13.

George B. Field, Cambridge, Massachusetts, and
Eric J. Chaisson, Haverford, Pennsylvania

Winter, 1984

* Astronomy and Astrophysics for the 1980s.
Volume 1: Report of the Astronomy Survey Committee
Volume 2: Reports of the Panels
Volume 3: Challenges to Astronomy and Astrophysics: Working Documents of the Astronomy Survey Commitee
(National Academy Press, 1982, Washington, D.C.)

Acknowledgments

Images for the color plates are provided by the following individuals and institutions:

Observers

J. Burns
D. Chance
P. Crane
J. Dickel
G. Dulk
G. Fazio
E. Feigelson
W. Forman
D. Gary
W. Gilmore
C. Jones
M. Morris
G. Neugebauer
I. de Pater
M. Reid
R. Schild
E. Schreier
E. Seaquist
C. Snyder
J. Stocke
R. Tresch-Fienberg
G. Withbroe
F. Yusef-Zadeh

Institutions

Columbia University
Harvard-Smithsonian Center for Astrophysics
Jet Propulsion Laboratory, California Institute of Technology
National Radio Astronomy Observatory
Space Telescope Science Institute
University of California, Los Angeles

Polaroid Corporation provided additional support in the reproduction of the color plates.

Quotations preceding each chapter were taken, with permission, from the following sources:

PREFACE: *Dialogues Concerning Two New Sciences* by Galileo Galilei, Leiden, 1638; translated by H. Crew and A. de Salvio, 1914, The Macmillan Co., New York.

CHAPTERS 1, 2, 4, 5: *The Sidereal Messenger: Unfolding Great and Marvelous Sights, and Proposing Them to the Attention of Every One, But Especially Philosophers and Astronomers* by Galileo Galilei, Venice, 1610; translated by E.F. Carlos, Dawsons of Pall Mall, London.

CHAPTERS 3, 6: *History and Demonstrations Concerning Sunspots and Their Phenomena* by Galileo Galilei, Rome, 1613; translated by S. Drake, 1957, Anchor, New York.

CHAPTER 7: *Dialogue on the Two Chief World Systems* by Galileo Galilei, Florence, 1632; translated by S. Drake, 1967, 2nd ed., Univ. Ca. Press, Berkeley and Los Angeles.

CHAPTER 8: *The Assayer* by Galileo Galilei, Rome, 1623; translated by S. Drake, 1957, Anchor, New York.

EPILOGUE: *Dioptrics* by Johannes Kepler, Augsburg, 1611; translated by E.F. Carlos, Dawsons of Pall Mall, London.

1
Radiation Visible and Invisible

Keys to the Universe

It would be altogether a waste of time to enumerate the number and importance of the benefits which this instrument may be expected to confer, when used by land or sea. But without paying attention to its use for terrestrial objects, I betook myself to observations of the heavenly bodies; and first of all, I viewed the Moon as near as if it was scarcely two semi-diameters of the Earth distant. After the Moon, I frequently observed other heavenly bodies, both fixed stars and planets, with incredible delight; and, when I saw their very great number, I began to consider about a method by which I might be able to measure their distances apart, and at length I found one. And here it is fitting that all who intend to turn their attention to observations of this kind should receive certain cautions. For, in the first place, it is absolutely necessary for them to prepare a most perfect telescope, one which will show very bright objects distinct and free from any mistiness, and will magnify them at least 400 times, for then it will show them as if only one-twentieth of their distance off. For unless the instrument be of such power, it will be in vain to attempt to view all the things which have been seen by me in the heavens, or which will be enumerated hereafter.

From *The Sidereal Messenger,*
by Galileo Galilei, Venice, 1610

WHAT DO WE MEAN by the invisible Universe? Simply those objects in space that emit types of electromagnetic radiation to which the human eye is insensitive. These radiations include radio waves, infrared and ultraviolet radiation, and x rays and gamma rays. Each of these different kinds of radiation travels at the velocity of light – 300,000 kilometers per second – just as does visible light itself, which is simply another kind of radiation.

Light is the most familiar kind of radiation to humans on planet Earth, as it enables us to see objects on the surface of our planet. But it also enables telescopes to detect objects deep in space. In fact, it was an optical telescope, or one sensitive to light, that Galileo first used to change forever the way that the oldest science – astronomy – is pursued. Ever since, optical astronomers have used telescopes sensitive to visible light to study the Universe.

In this book we purposely emphasize the accomplishments of those astronomers who work with invisible radiation, but not because the work of optical astronomers is unimportant. On the contrary, optical astronomy will doubtless continue to unlock secrets of the cosmos. But now our view of the Universe is undergoing a grand change. Within a single generation, by learning how to capture and analyze kinds of radiation other than light, we have discovered to our surprise that many of the most interesting phenomena – and even most of the matter – in the Universe do not emit light. Succinctly stated, most matter is invisible, and must be explored by using whatever kinds of radiation are emitted.

In this introductory chapter, we briefly review the various kinds of radiation, note some of the tools needed to detect and analyze it, and sketch some of the new ways used to image or display radiation in order to best extract information about the Universe.

The Electromagnetic Spectrum

Over the years, experiments have shown that electromagnetic radiation is a type of energy that moves at the velocity of light from one place to another. Radiation carries information, enabling us to find clues to the nature of the objects that emit it; thus radiation is the principal way that information travels in our Universe. It is the task of the astronomer to use the laws of physics to extract the information present in the radiation from cosmic sources, thereby advancing astronomical knowledge.

We usually think of radiation as a wave, but it is also possible to conceive of it as being composed of a stream of particles, called photons. According to the modern theory of atomic and subatomic phenomena termed quantum theory, radiation actually manifests both aspects; which aspect dominates in any given situation depends upon the way the observer interacts with the radiation. For example, when we see a television signal fading in and out, it may be the result of interference between the wave coming directly from the TV station, and another reflected from a passing plane, for interference phenomena are characteristic of waves. On the other hand, when a Geiger counter exposed to a source of x rays emits a "click", we know that it has registered an x-ray photon. All the different kinds of radiation share the attributes of both particles and waves.

An analogy might help. Imagine yourself to be in a canoe in the middle of a lake. There is no breeze, and the surface of the lake resembles a mirror. Moving your hand back and forth in the water, you will see a long chain of ripples moving out – a wave. But if you plunge your hand in and out just once, you will see a single pulse of ripples. This pulse is something like a particle, for it is localized in space and time.

That ordinary white light is made of various colors – red, orange, yellow, green, blue, and violet – can be easily proved by passing it through a simple prism, which separates the light into waves of different lengths.* Experiments have shown that red light has a wavelength of about 7×10^{-5} centimeter, nearly twice that of violet light, whose wavelength measures about 4×10^{-5} centimeter. The intermediate colors have wavelengths between these values, and together all the colors delineate a range termed the visible spectrum. (Optical astronomers often use another unit of wavelength, called the Angstrom, which by definition equals 10^{-8} centimeter; thus the visible spectrum spans 4000 to 7000 Angstroms.)

Light is the radiation to which human eyes are sensitive. As light enters our eye, the lens focuses it onto the retina, where it initiates small chemical reactions that send impulses to the brain. It is this complex chain of events we sense as light. Radiation having wavelengths outside the visible spectrum cannot be sensed this way, for our retinas are simply insensitive to it. At wavelengths larger than 7000 Angstroms lie the regimes of radio and infrared radiation; at wavelengths smaller than 4000 Angstroms are the domains of ultraviolet, x-ray, and gamma-ray radiation. Collectively, all

* The length of a wave is the distance between successive crests or successive troughs, just as in a water wave.

these spectral regions, including the visible spectrum, comprise what we call the electromagnetic spectrum.*

Consider the sketch (on page 7) of the electromagnetic spectrum, in which the wavelength scale is marked off in factors of ten. Note that the frequency at which the wave oscillates (measured in cycles per second, or Hertz) is inversely proportional to the wavelength. For example, a wave whose length is 1 centimeter has a frequency of 3×10^{10} Hertz, or 30 Gigahertz.

This figure displays the great range of wavelengths covered by the many different kinds of electromagnetic radiation. As shown, typical wavelengths extend from the size of mountains for radio waves, to the dimensions of atomic nuclei for gamma rays. Familiar objects are presented adjacent to the wavelength scale to help visualize the lengths of waves throughout the electromagnetic spectrum.

In designing this figure, we have carefully drawn the various spectral intervals to scale. Accordingly, we note the rather small size of the interval occupied by the visible spectrum, and thus begin to appreciate the fact that our eyes detect only a minute portion of the vast spectrum of electromagnetic radiation. Clearly, if objects in the Universe emit radiation outside the 4000–7000 Angstrom visible range – and they most certainly do – then the sheer size of the invisible domains suggests that a wealth of knowledge will likely be gained by detecting and interpreting such radiation.

We might well wonder why the range of visibility is so small compared to other parts of the electromagnetic spectrum. Why do human eyes respond to only a minute fraction of the many different kinds of radiation known to exist?

To answer this question, we must consider how radiation coming from space is affected by Earth's atmosphere, that is, the extent to which it is attenuated, or reduced in intensity, while passing through molecules of air. In the figure we have sketched a scale in which the shading is proportional to the amount of atmospheric attenuation, so that, for example, the atmosphere is completely opaque to radiation at those wavelengths where the shading is greatest; such radiation cannot penetrate our atmosphere even under the clearest weather conditions. By contrast, where there is no shading at all, our atmosphere is nearly transparent (except for the effects of clouds on light

* Incidentally, the various wavelength bands carry different names for radiation, including "rays", "radiation", and "waves". All these terms – and "photons" as well – refer to the same phenomenon of electromagnetic radiation, which as explained above has aspects both of waves and of particles. This is true for radiation of all wavelengths; the names are historical accidents.

radiation), meaning both that radiation from space can reach Earth's surface, and terrestrial radiation from man-made sources can escape unhindered into space.

What causes the atmospheric attenuation to vary along the spectrum? Why does visible radiation penetrate the atmosphere, while ultraviolet radiation does not? Why do relatively long-wavelength radio waves (those having wavelengths between about 1 and 1000 centimeters) pass unhindered through the atmosphere, while short-wavelength radio waves (about 0.1 centimeter wavelength) do not? Answers to these questions depend largely on the types of gases comprising Earth's atmosphere.

Atoms and molecules absorb radiation passing by, and do so more efficiently at some wavelengths than others. For example, molecular oxygen (O_2), a major constituent of our atmosphere, absorbs radio radiation having wavelengths less than about a centimeter. Similarly, water vapor and carbon dioxide (CO_2), important trace gases in Earth's atmosphere, are strong absorbers of infrared radiation, so that except for a few narrow bands (or "windows"), infrared radiation cannot penetrate our atmosphere. A third example involves the attenuation of visible light by atmospheric clouds; when the abundance of water vapor reaches high enough levels, clouds form – and with them rainy weather – thus preventing the direct passage of light from astronomical objects to our ground-based telescopes.

Ultraviolet, x-ray, and gamma-ray radiation is also completely blocked by Earth's atmosphere. At altitudes of about 50 kilometers (*i.e.*, about 150,000 feet), normal O_2 molecules, which are composed of two oxygen atoms, are broken apart into their individual atoms by such short-wavelength radiation emitted by the Sun. Almost immediately the free oxygen atoms recombine with remaining O_2 molecules to form a thin layer of ozone (O_3), which itself absorbs ultraviolet radiation. Thus, short-wavelength radiation trying to pass through our atmosphere is intercepted by O_2 or O_3 molecules, (as well as by molecular nitrogen, N_2).

The layer of ozone, by the way, is one of several insulating spheres that serve to protect life on Earth from the harsh realities of outer space. Not long ago, scientists judged space to be hostile to advanced life forms because of what is missing – breathable air and a warm environment. We now know that outer space is also harsh because of what is present – intense ultraviolet radiation and energetic particles from the Sun, both of which are injurious to health. The ozone layer is one of Earth's umbrellas; without it, advanced life could not exist.

Let us return to our earlier inquiry regarding the limited usefulness of the human eye. We see that in the figure, the visible spectrum is aligned almost exactly with a narrow domain or "window" in which the atmospheric

Continued on page 8

Types of Electromagnetic Radiation

Radiation is composed of electromagnetic waves. The different types of radiation are distinguished by their wavelength, which is the distance between successive wave crests or troughs. This diagram displays all the types of radiation observed by astronomers, arranged in order of wavelength, from 10^{-13} cm at the bottom to 10^6 cm at the top. The wavelength scale is logarithmic, each tick representing a factor of ten in wavelength. Wavelengths are given in centimeters at the left, and to the right they are given in the units appropriate for that wavelength range: kilometers for long radio waves, meters for intermediate radio waves, centimeters or millimeters for microwaves, microns (μ) for infrared, and Angstroms (Å) for visible and ultraviolet radiation as well as for x rays and gamma rays.

The next scale to the right gives the frequency corresponding to each wavelength; note that frequency is inversely proportional to wavelength. The standard unit of frequency measurement is Hertz (Hz), equivalent to the passage of one wave per second. Radio astronomers use kilo-Hertz (1 kHz = 1000 Hz), mega-Hertz (1 MHz = 1000 kHz), and giga-Herta (1 GHz = 1000 MHz), while in the visible, ultraviolet, x-ray, and gamma-ray regions it is customary to use energy units, electron volts (eV), instead of frequency units. (This is possible because according to a basic law of physics, the energy of a photon is proportional to the frequency of the wave.) X-ray astronomers use kiloelectron-Volts (keV) and gamma-ray astronomers use megaelectron-Volts (MeV).

The next column to the right indicates the transparency of the Earth's atmosphere at the corresponding wavelength, with the lighter shading corresponding to greater transparency. Until the 1950s, when astronomers began observing through the radio window, all astronomical observations were made through the visible window. Since the 1960s, the entire electromagnetic spectrum has been opened up for observation by telescopes orbiting the Earth above the atmosphere.

WAVELENGTH (cm)		FREQUENCY	ATMOSPHERIC TRANSPARENCY	TYPE OF RADIATION		OBJECT
10^{6}	10 km	30 kHz				
10^{5}	1 km	300 kHz		AM		Mountain
10^{4}	100 m	3 MHz		Radio waves		
10^{3}	10 m	30 MHz				Office building
10^{2}	1 m	300 MHz		FM; VHF		Human
10^{1}	10 cm	3 GHz		UHF		
1	1 cm	30 GHz		Microwaves		Thumb
10^{-1}	1 mm	300 GHz				Pinhead
10^{-2}	100μ			Far infrared		
10^{-3}	10μ	PHOTON ENERGY		Infrared	*red*	Housedust
10^{-4}	1μ	1.2 eV		Visible light	*orange yellow green blue violet*	
10^{-5}	10^{3}Å	12 eV		Ultraviolet		Bacterium
10^{-6}	100 Å	120 eV		Extreme UV		
10^{-7}	10 Å	1.2 keV		Soft x rays		Virus
10^{-8}	1 Å	12 keV		x rays		Atom
10^{-9}	0.1 Å	120 keV				
10^{-10}	0.01 Å	1.2 MeV				
10^{-11}	10^{-3} Å	12 MeV				
10^{-12}	10^{-4} Å	120 MeV		Gamma rays		
10^{-13}	10^{-5} Å	1.2 GeV				Atomic nucleus

attenuation reaches minimum values, at wavelengths between 3 and 9×10^{-5} centimeter or 3000 to 9000 Angstroms. Accordingly, while most of the Sun's infrared radiation and all of its ultraviolet radiation are blocked by our atmosphere, its visible radiation pours right in. It is no coincidence that the human eye has its greatest sensitivity in this window; no doubt our eyes have physiologically evolved over millions of years so as to use efficiently the most intense radiation reaching Earth's surface from the Sun, thereby allowing us to identify objects and to move about effectively on the surface of our planet. No wonder that we give this type of radiation a special name – "light".

If our eyes, for some reason, responded only to infrared radiation, they would be much less useful to us. The Sun's infrared radiation cannot penetrate our atmosphere very well, and thus could at best only dimly illuminate objects on Earth's surface. Even worse, if our eyes were sensitive only to ultraviolet radiation or to x rays, they would be useless, since neither type of radiation reaches Earth's surface.

Why haven't human eyes evolved to be also sensitive to radio radiation? After all, our figure shows that Earth's atmosphere is largely transparent to most radio radiation. Why can't we perceive radio waves, or at least use them to judge the position of objects around us? The answer is probably that, because of its high surface temperature (about 6000 degrees Celsius), the Sun emits visible radiation much more strongly than radio radiation. Had there been a source of radio radiation more intense than our Sun's visible radiation, then life forms might have learned to perceive objects on the surface of Earth by picking up the radio waves reflected by them, instead of by looking at the reflections of visible light. Even if they had, however, their "radio eyes" would have been able to achieve only a fuzzy image, for a law of physics decrees, as we shall see, that short-wavelength radiation is required to resolve small details.

Tools of Astronomy

Knowledge of the cosmos usually advances in three phases: collection and detection of radiation from space, storage of the resulting data, and analysis of that data. The first phase is normally accomplished with a telescope of some sort, the second might employ photographic film or magnetic tape, while the third requires us to apply the laws of physics to create a mental image, or model, that explains the acquired data. Of course, theoretical work plays an important role, and often suggests the acquisition of crucial data, but, more often than not for an empirical science like astronomy, observations of cosmic phenomena precede their theoretical prediction.

In the case of optical astronomy, a concave mirror acts to collect light and to focus it at a point where a detector records its presence. The detector could be a human eyeball, but to obtain a permanent objective record, astronomers use photographic plates or electronic detectors that record data on computerized magnetic disks. Telescopes are often made more sophisticated by inserting, in front of the detector, devices such as filters, prisms, or spectrographs to help extract information about the spread of light over its component wavelengths.

The largest astronomical reflector currently operating is the 6-meter diameter telescope atop the Caucasus Mountains in the Soviet Union. A 5-meter (200-inch) mirror sits atop Mount Palomar in California, while several 4-meter telescopes now adorn mountain summits in Arizona, Hawaii, Australia and Chile. Legions of smaller telescopes also scan the skies nightly, contributing much to our knowledge of the cosmos.

A radically different type of optical telescope was recently built on Mt. Hopkins, in Arizona, a few miles from Tucson. Called the *Multiple Mirror Telescope*, the primary collector of this device is not a single, huge, and heavy mirror, but rather six smaller ones, each 1.8 meters (72 inches) in diameter. These mirrors work in tandem to approximate the capabilities of a single mirror 4.5 meters in diameter. As we shall see below, this novel design may well spawn a new generation of large telescopes in the years ahead.

Large telescopes are usually more advantageous than small ones for two reasons. First, larger telescopes collect light over a greater area and can thus bring more radiation to a focus where it can be studied. Second, larger telescopes are in principle capable of finer angular resolution; that is, they can better distinguish, not only closely adjacent objects of similar brightness, but also the fine details of extended objects.

Even our largest telescopes have their limitations, however. For example, although the 5-meter telescope on Mount Palomar is theoretically able to attain an angular resolution as fine as 0.02 arc second (that is, distinguish between two objects that close together in the sky), in practice it cannot do better than about 1 arc second*. As a matter of fact, no ground-based optical telescope can resolve cosmic objects much closer together than one arc second because the random motions of air tend to blur the paths of light waves passing through our atmosphere. (These are the motions that cause the stars to twinkle in the evening sky.)

* An arc second is a unit used to measure small angles; it equals 1/60th of an arc minute, which in turn is 1/60th of an arc degree. Whereas there are 360 arc degrees in a circle on the sky, one arc second is the angle subtended by an American dime when viewed from a distance of about 3 kilometers.

Optical telescopes in orbit about the Earth or emplaced on the Moon can overcome this limitation. Without atmospheric blurring, extremely fine resolution can be achieved, subject only to the engineering limitations of building large structures in space. For example, the *Space Telescope*, scheduled to be launched into Earth orbit by NASA's *Space Shuttle* in 1986, should be able to resolve objects with an accuracy as fine as 0.05 arc seconds, even though its 2.4-meter mirror is not as large as the largest telescopes on the ground. Thus, this orbiting observatory, to which we shall refer throughout our book, should overnight give us a view of the Universe some twenty times sharper than ever before available.

In contrast, radio telescopes cannot usually resolve objects from one another as well as do their optical counterparts. Despite their huge sizes, typically spanning tens or even hundreds of meters, radio telescopes generally have poor angular resolution. This is because they are sensitive to radiation whose wavelengths are millions of times larger than those of light, and according to the laws of physics, these longer wavelengths impose a corresponding crudeness in angular resolution. While optical telescopes resemble human eyes, radio telescopes in this respect are more like human ears, giving only an indication of the direction from which the radiation is coming. The analogy is apt in another respect also. Eyes receive all visible wavelengths, and record instantaneously the sensation of color. Ears, however, perceive different wavelengths separately (though we can hold many musical tones in our consciousness). In a similar way, radio telescopes normally receive radiation within only a narrow band of wavelengths; to detect radiation at another radio wavelength, we must retune the equipment (much as we tune our TV sets to different channels.)

The best angular resolution of a single radio telescope is about 10 arc seconds, an order-of-magnitude coarser than the 1 arc second possible with the largest optical telescopes. Confronted by this difficulty, radio astronomers have devised ways to achieve much higher resolution by using two or more radio telescopes simultaneously to form a so-called "interferometer". When several radio telescopes work in tandem to observe the same object at the same wavelength and the same time, they are equivalent as regards angular resolution to a huge, imaginary telescope whose diameter equals the distance (or "baseline") separating the smaller, individual telescopes. In this way, resolution of a few arc seconds can be achieved at typical radio wavelengths (say 10 centimeters) by using two telescopes separated by 5 kilometers. The most sensitive radio "telescope" in the world is just such an interferometer, a collection of 27 telescopes spread across a baseline of nearly 30 kilometers on the Plains of San Augustin in New Mexico. Called the *Very*

Large Array, this instrument is one of the daily workhorses of contemporary astronomy, and is contributing much to studies of the fine-scale structure of many kinds of cosmic objects.

The finest resolution achieved by radio astronomers to date – about 0.001 arc second – was recently attained with an even larger interferometer, one whose baseline stretched nearly across our entire planet. Known by the tongue-twisting name of "very long baseline interferometry", this international technique employs radio telescopes in North America, Canada, Europe, and the Soviet Union to achieve angular resolutions about a thousand times better than the best results of conventional optical astronomy. Significantly, then, angular resolution in radio astronomy is now limited not by the atmosphere of our planet but by the fact that the entire Earth is only so big (12,800 kilometers in diameter).

The principal advantage of radio astronomy is that it permits us to perceive radio-emitting objects even if they are dark. Whether objects are intrinsically cool and thus devoid of any light, or hidden behind clouds of dust that obscure the view of optical astronomers, this long-wavelength science enables radio astronomers to probe the invisible regions of space where there is literally nothing to see. This is possible because even cool objects emit radio radiation, since physics teaches us that the colder the object, the longer is the wavelength of the emitted radiation. Moreover, we know of objects in space, such as pulsars, which are extremely hot in the sense that they contain particles moving at nearly the speed of light, but which emit far more radio waves than they do light. Needless to say, radio astronomers were delighted to encounter such objects; indeed, they are still best studied by their radio emissions. And of great importance, radio radiation can propagate through dust clouds in space, just as radio signals on Earth can penetrate cloudy or foggy weather, thereby enabling radio astronomers to construct images of regions completely hidden from view.

Infrared techniques provide another way to observe objects and regions that are inherently dark. Warm objects – those not so cool as to emit only radio waves, yet not hot enough to emit an appreciable amount of light – emit radiation having wavelengths intermediate to those of radio and light waves. Because infrared radiation, somewhat like radio radiation, has wavelengths larger than the typical size of dust particles in space, much of it can make its way through interstellar dust clouds that otherwise block the passage of shorter-wavelength light. Since the atmospheric attenuation is not complete in certain parts of the infrared domain, some important observing is done from the ground, particularly in narrow wavelength bands or "windows" in which the atmospheric attenuation is much reduced at wavelengths between 1 and about 20 microns (10^{-4} centimeters). Because of some residual attenuation at infrared wavelengths, however, the work is increasingly carried out by infrared detectors either hoisted above most of our atmosphere by high-flying balloons, rockets, and aircraft, or placed into Earth orbit above our entire atmosphere.

For example, the recently launched *Infrared Astronomical Satellite* is contributing much to our knowledge of the clouds of galactic matter that seem destined to become stars – regions composed of warm gas that can be neither seen with optical telescopes nor adequately studied with radio telescopes. The primary device aboard this U.S.-Dutch-British satellite is a 0.6-meter mirror that focuses the incoming infrared radiation with an angular resolution as fine as 30 arc seconds. Even as this is being written, there is a report that this satellite has just discovered not just one, but several stars around which particles are orbiting. Not planets, apparently, but particles from millimeters to meters in size, which could well serve as the first stage of the accumulation of material to form planets around a star far beyond our own Solar System.

To the short-wavelength side of the visible spectrum lie the ultraviolet, x-ray and gamma-ray domains. Since our atmosphere is totally opaque at wavelengths shorter than about 3000 Angstroms, no ultraviolet or x-ray observing can be done from the ground, not even from the highest mountaintop. Rockets, balloons, or satellites are an absolute necessity. Throughout the past decade, a number of ultraviolet, x-ray, and gamma-ray facilities have been orbited to sample this short-wavelength radiation, and much useful knowledge has been gained, yet currently all but two of these satellites either have run out of the gas needed to orient themselves or have scanned the skies to the limits of their sensitivities.

Perhaps the most noteworthy of these satellites has been the *High Energy Astronomical Observatory* # 2, named the *Einstein Observatory* in honor of the birth-centenary of the great scientist in 1979, the year the satellite began scientific work. This craft has on board equipment capable of detecting cosmic x rays, but it does so in a way that differs considerably from the devices used to detect other kinds of radiation. Curved surfaces cannot be made smooth enough to reflect radiation of such small wavelength, often less than 100 Angstroms; x rays normally go straight through lenses or mirrors, just as they do when used to x ray the human body. Instead, incoming cosmic x rays graze along a series of variably curved and highly polished metal surfaces that gently guide them to a focus where they are detected. In this way, x-ray images of cosmic sources are collected by satellites above our opaque atmosphere, converted into radio signals by onboard equipment, transmitted through the atmosphere and received by conventional radio telescopes on the ground, and finally transformed into electronic signals that can be viewed and analyzed on a video screen. Although the collecting diameter of the *Einstein Observatory* is only 0.6 meter, the short wavelength of the x rays enables its spatial resolution to be a fine 3 arc seconds. Detecting x rays from objects which are so hot that almost all of their radiation is in x rays, not light, this satellite has done for x-ray astronomy what Galileo's small telescope once did for optical astronomy.

Clearly, astronomy has come of age in nearly every part of the electromagnetic spectrum. All our equipment, on the ground and in orbit, comprises a veritable arsenal of devices available to modern astronomers. The electromagnetic spectrum is so large that it would be unrealistic to expect that we could use any one apparatus to detect radiation of all wavelengths. Different tools address different objectives, and some do the job better than others. In the end, they all supplement one another, helping to accumulate an ever-growing store of basic astronomical knowledge about our richly endowed Universe.

Astronomical Image Processing

The computer revolution that has changed all of our lives has also changed the basic nature of astrophysical research. Theorists can use computers to simulate physical events – like explosions of stars or collisions of galaxies – in much the same way that moviemakers simulate flights of spaceships. The theorist puts the laws of physics into the simulation and, by comparing the results with observations, tests whether the proposed model is correct. For the observer, the changes introduced by computers are just as dramatic.

Astronomers have traditionally recorded images on photographic plates. Many thousands of detailed photographs of the night sky have been accumulated and stored in archives where they are still used in scientific research. However, the medium of photography is inherently limited. Most important, photographic film is much less sensitive than available electronic detectors, making photographic exposure times much longer by comparison. Although photographs can show a great deal of detail, they are incapable of simultaneously recording the large range of intensities inherent in astronomical images; thus, a single photographic exposure cannot record both the faintest features and the brightest features of an object. Also, the complex response of film at different light levels and wavelengths must be taken into account. Furthermore, film is awkward to use in remote-sensing applications; for example, the tremendously exciting images obtained by the planetary probes of the last decade would not have been possible without the use of modern electronic detectors whose output can be radioed back to Earth. Another very important limitation of photographic images is that the information stored on them is not readily accepted by modern digital computers.

How do astronomers record images today? There are many answers to this question, but virtually all of them involve the production of computer-compatible data in digital form – a string of numbers. For example, in the visible and infrared regimes, new light-sensitive silicon chips called CCD's (for "charge-coupled devices") and CID's (for "charge-intensified devices"), originally developed for military or commercial applications, are paving the way for the new astronomy. These chips produce electrons when a photon of light hits them, after which the electrons are stored on the chip itself. Periodically the numbers of electrons are counted electronically, converted into digital counts, and stored by a computer on magnetic disk or tape. CCD's and CID's have the advantage over photographic film in that they can record bright and faint objects simultaneously. In quantitative terms, a photographic film typically is limited to about a factor of 100 between the lowest and highest intensity it can record, whereas a CCD can accommodate a factor of more than 10,000; a CID is limited only by the computer itself – usually to a factor of a billion or so.

Radio astronomers collect radiation in a rather different way, using large radio telescopes or groups of telescopes to map the sky. Likewise, x-ray astronomers use special detectors designed to be sensitive only to x rays. In either case, the outputs of all these detectors are numbers – billions of them – which, when properly arranged into a two-dimensional array, and processed and displayed with special equipment, form images that can be examined, interpreted, and manipulated for maximum scientific usefulness. Unless otherwise identified, all the illustrations of astronomical objects in this book originated with digital images that have been processed using a computer.

Once the image is represented as an array of numbers in the computer, all the analytical tools at the programmer's command may be brought to bear. As far as the astronomer is concerned at this point, it is irrelevant whether the data forms a radio, infrared, optical, ultraviolet, or x-ray image. The same processing techniques may be applied to all these kinds of data regardless of their origin. As we shall see in this section, special equipment and computer programs have become available in the last few years that enable us to extract large amounts of information from digital images.

Let us consider briefly the sheer volume of data being produced by astronomical research programs today. The "floppy disk" used for storage with personal computers can typically store about ⅓ of a Megabyte (1 Megabyte equals a million characters or about a half-million digits); typical "hard disks" used with personal computers can hold about 10 Megabytes. One astronomical image made by a CCD requires about ½ Megabyte, and a typical night of observing at a telescope will produce from 10 to 50 images, or 5 to 25 Megabytes of data – the equivalent of 15 to 75 floppy disks.

The information contained in a single photograph can be converted to digital form by scanning it slowly with a special device called a photodensitometer, which records the amount of intensity within each tiny area of the photograph. The resulting data requires up to 1000 Megabytes of storage. A radio experiment on the *Very Large Array* typically produces images having at least 10 Megabytes, while x-ray images from the *Einstein* satellite are often in the range of 1 to 10 Megabytes each.

If we estimate the total amount of data represented by all the CCD and CID images collected by a single observatory in a single year, we expect about 1000 images, or approximately 500 Megabytes of data. The total data base produced each year by the *Very Large Array* amounts to 1000 to 10,000 Megabytes. In its short three-year lifetime, the *Einstein* x-ray satellite produced more than 25,000 images, thus yielding a total data base of more than 10,000 Megabytes.

Clearly, this flood of data is too large to be stored on easily accessible floppy disks; rather, most of the data are stored on thousands of magnetic tapes. In the very near future, laser disks similar to those currently available to consumers for home video entertainment will become an alternative to magnetic tape. A single laser disk can record at least 1000 Megabytes of information, and thus a modest number of such disks should suffice to store the entire output of an optical, radio, or x-ray observatory for a year.

A brief comparison of the speeds of various computers may provide a perspective on these tools required to perform image processing. For convenience, the astronomer thinks of an image as a numbered array of individual, small areas called "pixels" (for picture elements). An astronomical image usually has from 500 to 2000 rows and columns of pixels. Some personal computers having graphics capability can display an image having a few hundred by a few hundred pixels; typically each pixel can display 8 to 16 different colors. With such a personal computer, we can add two 512 x 512 images in about ten minutes. This would be useful if two exposures were made of the same object, and we wanted all the information available in both exposures.

By comparison, a small mainframe or mini-computer used for scientific calculations can do the same job in a few seconds. However, these specialized computers do not usually have the ability to display images unless specific equipment has been purchased for the purpose. If it has, truly astonishing capabilities are available, for then we can obtain up to 256 colors or shades of gray in each pixel. Because multiple images may be displayed simultaneously, such computers provide a total color palette of over 16 million colors!

These days, most astronomical observatories are equipped with special computer and display equipment known as image processing systems. Such systems can display an array of numbers stored in the computer as an image on a high-resolution color television monitor having between 512 x 512 and 1024 x 1024 pixels. Such systems permit simple operations to be performed on an entire image in 1/30th of a second or less. Because these systems are so fast, astronomers can actually manipulate images while sitting at the observing console, changing the desired parameters at will, and watching the results appear almost instantly on a color television.

Advanced image processing systems offer an amazing range of capabilities for manipulating the data in astronomical images. Astronomers now interact directly with the data, thus permitting them to achieve the precise results desired. For example, one of the most common uses of an image processing system requires us to manipulate the relative intensities in various parts of an image; such "massaging" can highlight certain features of interest in much the same way that photographers can enhance certain features by controlling the development of negatives or adjusting the exposure of prints. And because most digital astronomical images have a wider range of intensities than the TV can display, the ability of the image processing system to display the low or high intensity areas of the image enables the astronomer to examine any feature of interest.

Here is an example of image processing used to extract scientific results from raw digital images. Figure 1a shows an optical CCD image of a typical galaxy, NGC 7479. In addition to the central structure of the galaxy, we can also see some parts of its outlying emissions. In the background, around the galaxy proper, we see virtually nothing.

Figure 1b shows the same image after the contrast has been enhanced to highlight the detail in the outer parts of the galaxy; details are also now visible in the background. Note, however, that the central structure of the galaxy is now "burned out"; no details are visible. This is a typical problem plaguing all astronomical imaging, because the actual range of intensities is so large. Fortunately, with an image processing system, astronomers can choose whether they wish to study the high-intensity part of the image (such as the central region of the galaxy in Figure 1a) or the low-intensity part (such as the outer regions of the galaxy in Figure 1b). By manipulating the controls of the system, we can watch the image change before us on the screen.

This single example points up an important advantage of image processing systems – the astronomer can interact intimately with the data. We can use our personal abilities to deal with images in order to complement those of the image processing system. After all, the combination of the human eye and brain can recognize and interpret features in ways that are often difficult or time consuming for a computer, and even occasionally spot interesting new features that might otherwise be missed.

Naturally, humans learn how to interpret data more effectively as each new technique is applied to image processing. Can computers participate in the learning process? Today most computers cannot learn but merely execute pre-defined programs. Even so, a new class of computers and programs called "expert systems" are now beginning to demonstrate an elementary ability to learn. The essence of such systems is that they are programmed to help the user according to predefined rules provided by a human expert. For astronomers interested in classifying galaxies into various types, for example, such systems might assist by learning to avoid certain errors commonly made in assigning galaxy types.

Another simple but important task that today's image processing systems can accomplish is to use the numbers stored in the computer to calculate important physical properties of the objects in the image. For example, an astronomer might be examining the image of the Ring Nebula shown in Figure 2a. This so-called planetary nebula is a diffuse gaseous envelope surrounding a star that expelled matter at some time in the past. By moving a small cursor – a square or circular indicator on the screen – the astronomer may define a region of interest within the image also shown in Figure 2a. Within this region, the total intensity of radiation, measured by the number of electrons counted by the CCD, the mean number of electrons counted in each pixel, the minimum and maximum number of electrons counted, and the spread of the number of counts around the mean value, can be computed almost instantaneously at the push of a button. Such numerical information is critically important to publish, both so other observers can check the data, and so theoretical models of the object can be compared with it.

The astronomer can also select a line through the image and plot the intensity of radiation along this line. Figure 2b illustrates such a plot of intensity along a line through the gaseous envelope and the central star. Note how the plot numerically displays the relative intensities of the star, the gaseous envelope, and the background, thereby reducing to numbers a range of intensities that is difficult to judge accurately by merely examining the relative shading of the image. The plot shows the location of each pixel along the bottom of the image and the intensity in each pixel along the side of the image. Notice that the diffuse envelope is more intense than the surrounding background and that the central star is brighter still. Since this is a

CCD image, the light intensity is measured in terms of electron counts; that is, the number of electrons produced when the light from the nebula was focussed by the telescope onto the CCD chip. Because the behavior of the CCD is well understood, we can convert the numbers of electrons to the numbers of photons actually hitting the CCD and thus to a measurement of the intensity of the object itself.

Most astronomical images are recorded using a single band of wavelengths, or color. Several "monochromatic" images of the same object can be combined in the image processing system to yield, in effect, a four-dimensional image that includes two spatial dimensions, along with intensity and wavelength. Thus, each pixel is labelled by four numbers, as we shall see shortly. Another, simpler use of color, called "pseudocolor," takes advantage of the fact that the eye is often more sensitive to changes in judicious choices of colors than to shades of intensity. For example, figure 4 shows a pseudocolored image of the planetary nebula NGC 7027. Each color corresponds to a specific intensity value. In this representation, color represents intensity, not wavelength; hence the term "pseudocolor."

Sometimes it is useful to record radiation of different wavelengths from the same object by making several exposures of the object through filters that eliminate all but certain wavelengths of radiation. Such multiple-wavelength images of the same object can then be displayed as a single three-color illustration, usually called a "false-color" image. For example, Figures 3b, 3c, and 3d show three CCD images of the galaxy M51 taken at three different wavelengths, blue/green, red, and infrared, respectively. When these images are combined in the image processing system, the resulting image, Figure 3e, shows a complex and subtle coloration owing to the fact that various parts of M51 are emitting radiation of the three wavelengths with different relative intensities. The pinker areas are sites of new young stars, whereas the dark brownish areas highlight lanes of interstellar dust and old stars. Notice that some of the individual stars in the image are much redder than others; such information at once indicates to the astronomer that these stars are worthy of further study.

We can even subtract one image from another to extract further information. In this case just discussed, we can thereby enhance the dust lanes and sites of star formation. For example, in Figure 3f, the infrared image has been subtracted from the blue/green image and the result pseudo-colored to show the dust lanes in blue/green and the new star sites as white.

Continued on page 23

(1a)

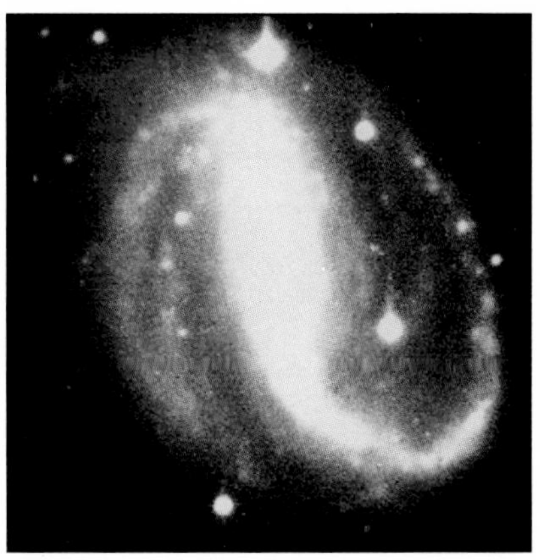

(1b)

Figure 1 NGC 7479, a spiral galaxy observed at visible wavelengths. These images were generated by computer to highlight the central region of the galaxy *(a)* and to bring out details in the spiral arms *(b)*. They were captured with a digital camera specifically designed for astronomical applications. It uses a light-sensitive silicon chip known as a charge coupled device, or CCD, which is much more sensitive to light than conventional film and which produces images that are computer-compatible. The chip is cooled using liquid nitrogen to reduce inherent system "noise," which would swamp the faint images that the camera is designed to detect.

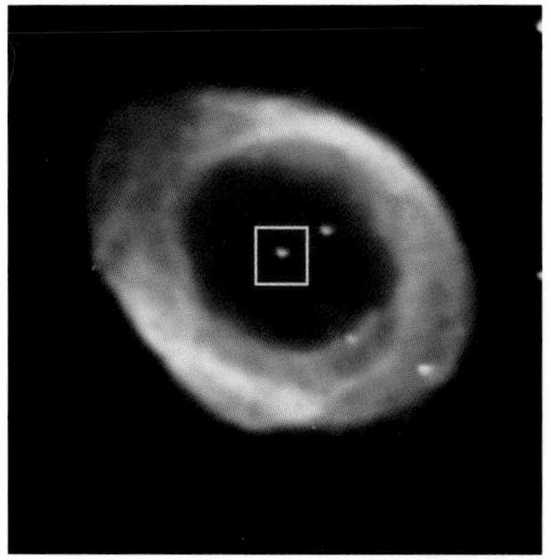

(2a)

Figure 2 The Ring Nebula in Lyra, a planetary nebula in the Milky Way galaxy, shown here *(a)* to highlight the wisps of gas that form the ring of material ejected when the central star exploded over 20,000 years ago. The "box" cursor in the center of the image may be positioned by the astronomer anywhere in the image and is used to extract data from the image, such as the total light gathered and the average density per picture element, or pixel. A computer-generated "intensity profile" of the Ring Nebula *(b)* through the stars and nebula. The resulting plot shows the relative brightness of the image at each point along the cut. Both of these processed images were generated from a single CCD source image.

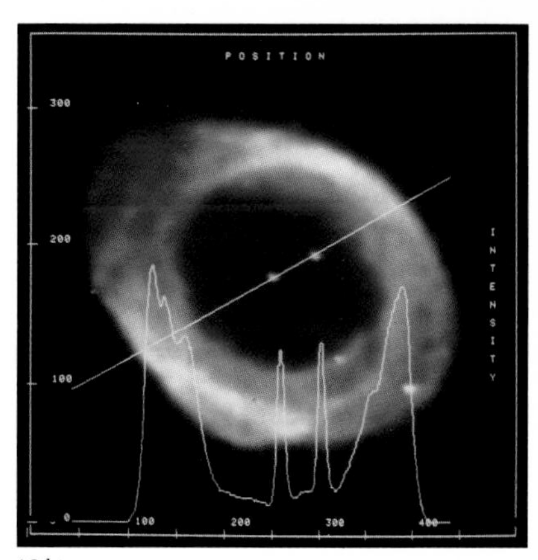

(2b)

Optical images recorded with a CCD camera on Polaroid Polacolor ER Land Film Type 809.

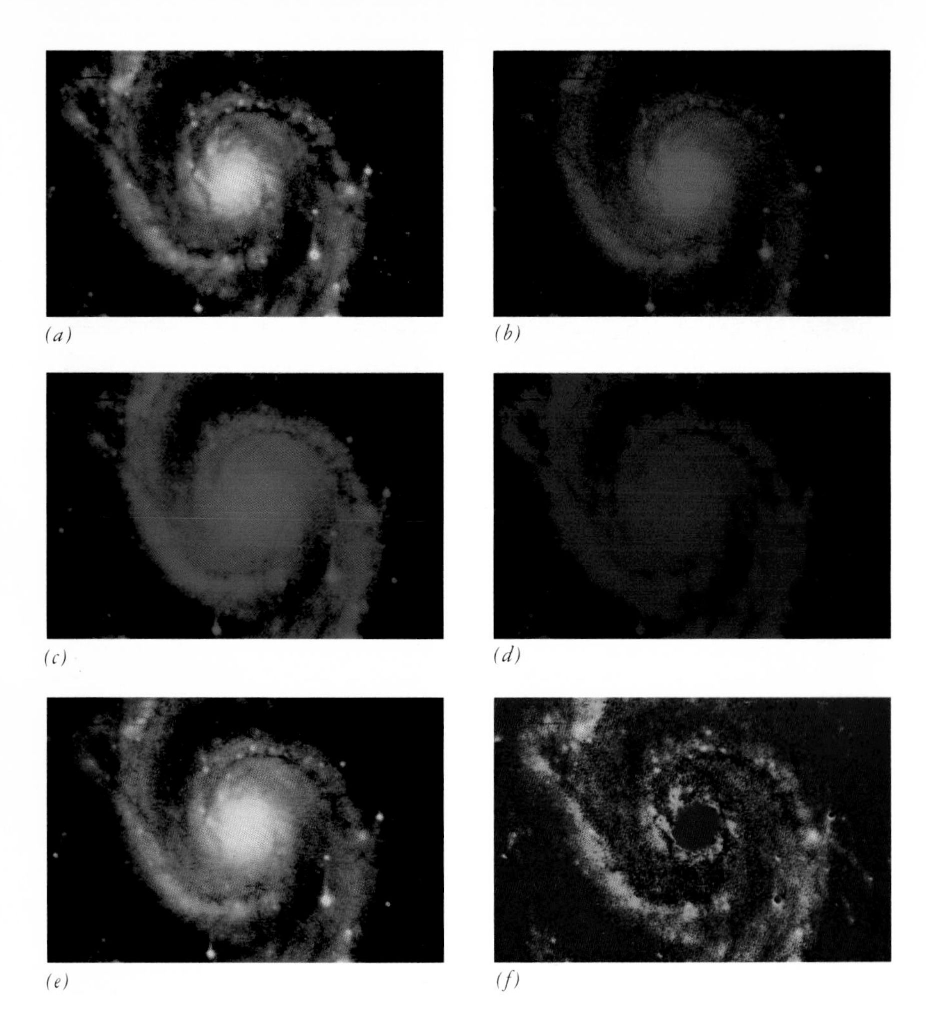

Figure 3 M51, a galaxy whose evolution has been affected by gravitational interactions with a companion galaxy NGC 5195 (not shown), resulting in areas of new star formation along the spiral arms. Figure *(a)* shows a monochromatic view of this galaxy. Figures *(b)–(d)* show exposures of the same galaxy in three wavelength bands using blue/green, red, and infrared filters (shown in blue, green, and red, respectively). Figure *(e)* shows the result of adding *(b)–(d)* to produce a single "multispectral image." The pinker areas are sites of new star formation and the brown, shadowy areas are sites of older stars and dust lanes. Figure *(f)* is the result of subtracting the infrared image from the blue/green image, yielding an enhanced view of features in figure *(e)*. All of these processed CCD images were created using the three source images *(b)–(d)*.

Optical images recorded with a CCD camera on Polaroid Polacolor ER Land Film Type 809.

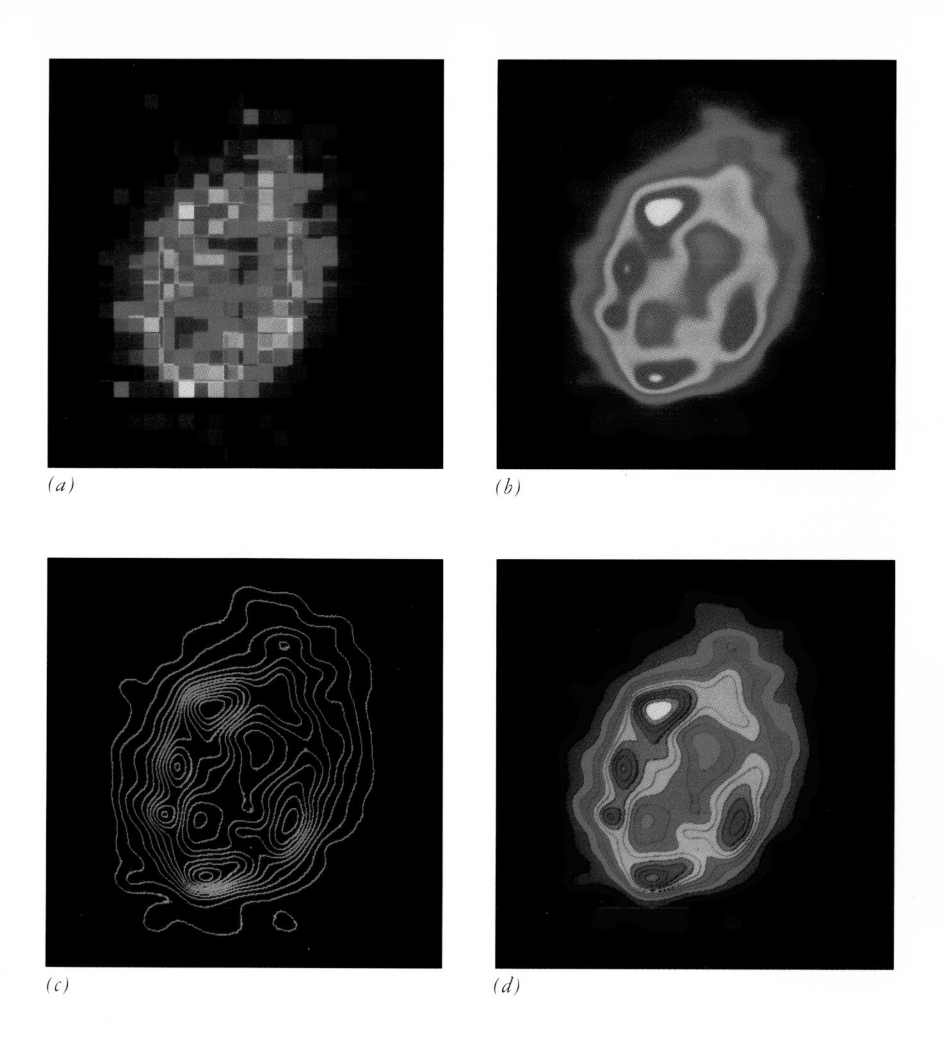

Figure 4 NGC 7027, a planetary nebula observed at infrared wavelengths. Each square in figure *(a)* corresponds to a single picture element or pixel. Using an image processor, this low-resolution image is manipulated by a mathematical technique known as gaussian convolution to produce the smoothed image shown in figure *(b)*. Figure *(c)* shows a contour map derived from the smoothed image. Figure *(d)* shows the result of superimposing the contoured color levels on the smoothed image. The image was obtained with an infrared digital camera based upon the technology known as charge injection devices, or CID's.

Infrared images recorded with a CID camera on Polaroid Polacolor ER Land Film Type 809.

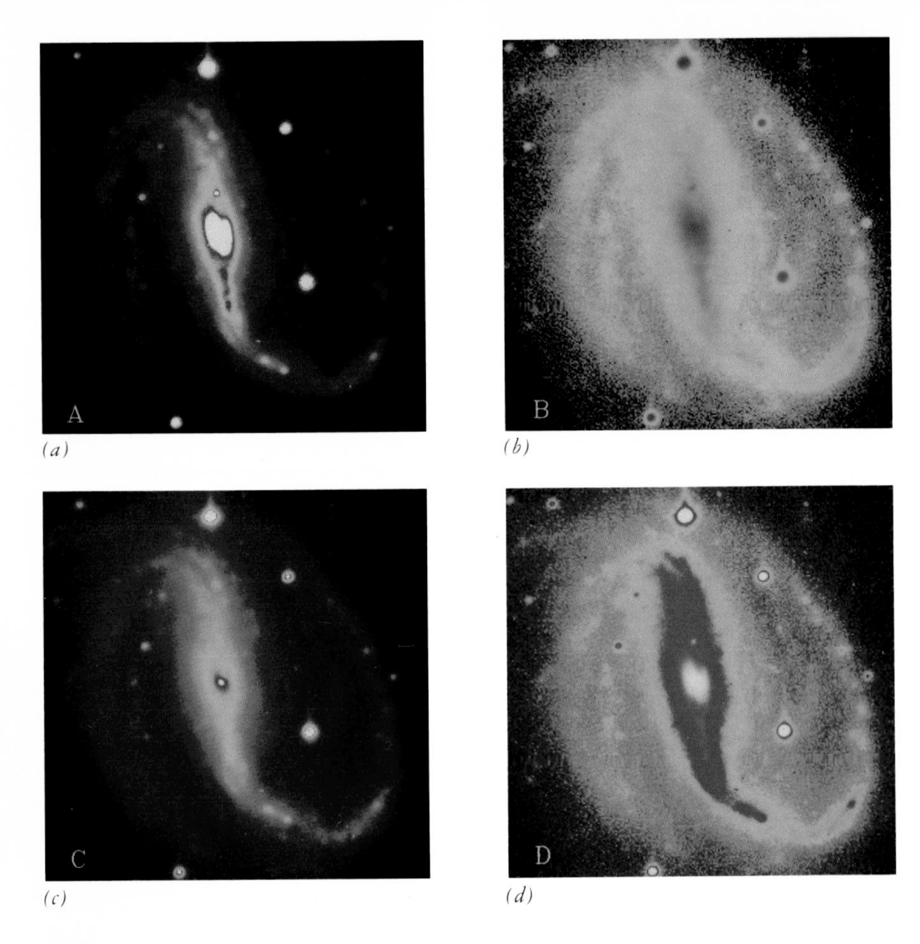

(*a*) (*b*) (*c*) (*d*)

Figure 5 NGC 7479, a spiral galaxy shown here with four different image processing approaches designed to extract information in the image. Figure (*a*) shows the galaxy with a linear transformation, figure (*b*) with a logarithmic transformation, and figures (*c*) and (*d*) with histogram equalization. Each of these transformations is designed to bring out information that may be hidden in the image; they provide, in a sense, a variety of ways of looking at the same thing. The linear transformation presents the data in the same manner that the CCD camera captures it. It also permits astronomers to look for many features that may be hidden in the data. It does not permit one to easily enhance both fainter and brighter features simultaneously. The logarithmic transformation lets one view the image as the eye does. It "compresses" the information in the image so that each level of grey is twice as compressed as the preceding level. Histogram equalization is a powerful technique that distributes the information in the image equally over the number of grey levels. This technique is intended to equally enhance all features contained in the image.

Optical images recorded with a CCD camera on Polaroid Polacolor ER Land Film Type 809.

Figure 4a shows an original infrared image of the planetary nebula, NGC 7027. This image was observed with an infrared CID having a field of view of about 16 x 16 pixels. With special features in the image processing system, this image has been magnified by a factor of eight in Figure 4a, making the individual pixels appear as discrete blocks. Such a display is useful in representing the raw data without any additional processing steps which could introduce errors in the image. However, such a display can be tricky to interpret, for it is hard to locate precisely the individual infrared sources and to determine their shape and extent.

Fortunately, a number of techniques can alleviate this problem. One such technique is to average between the pixels to get a much smoother-looking image. Many other techniques are available for smoothing data; we show one known as "gaussian convolution" in Figure 4b. This image no longer shows the hard-edged square pixels of Figure 4a but appears fuzzy and out of focus. This is not misleading because the original image does not contain much information, and any processed image seeming to show new structure would be highly suspect.

Nonetheless, it is often desireable to sharpen such an image to simplify the interpretation of the data. Again the image processing system comes to the rescue. While the processing in Figure 4b smoothed the image, we can now sharpen it by restricting ourselves to 8 or 10 colors. This results in well-defined boundaries between the colors. Figure 4d shows this doctored image, wherein each boundary represents a curve of equal intensity. Now the infrared sources, their shapes, and even their relative intensities are much more apparent. If desired, the image itself can be removed, leaving behind just the intensity map shown in Figure 4c. This is a useful step when preparing images for publication in scientific journals where color illustrations are prohibitively expensive.

Be sure to realize that the entire production of Figures 4 a-d took less than a minute of computer time on the image processor. The procedures used were limited not by the equipment but by the imagination of the user.

As a final example of image processing techniques, consider a problem that arises time and again, namely successfully displaying the full range of intensities in an image in a single illustration. A variety of techniques can accomplish this, the simplest of which is to arrange the intensities by factors of two. Figure 5b again shows the galaxy NGC 7479, this time with such a transformation. This tends to suppress the brightest parts of the image while enhancing the background. However, this correction is often too severe for many images and too much information is lost in the center of the image.

Another method for presenting a wide range of intensities is to distribute the information in the image equally over the available number of gray levels or colors. To achieve this goal, we count the number of pixels at each possible intensity value. The data are then reassembled in such a way that each grey level or color contains about the same number of pixels. Fortunately, many image processing systems have special equipment to accomplish these tasks. As is apparent in Figures 5c and 5d, this image shows more features of the galaxy than the previous examples. A recently developed variation of this technique allows the user to specify interactively how the information in an image is to be distributed over the available shades or colors. In this way, we can enhance to the desired degree any faint details in the astronomical image. Another technique, called "local area histogram equalization," adapts the amount of enhancement to the different conditions in the different parts of the image.

The examples described above give merely a glimpse at the rapidly burgeoning field of digital image processing. Many more techniques exist to extract information from images, and new ones are constantly being developed. The increasing availability of personal computers promises to bring many of the capabilities available now only in research laboratories into offices, factories, and ultimately our homes. To be sure, image processing is rapidly taking its place an an indispensible tool of our civilization.

Interstellar Ammonia

What interstellar molecules should one look for? Charles Townes, a Nobel Laureate in physics for his discovery of microwave amplification by stimulated emission of radiation (MASER), had developed techniques for searching for such molecules in interstellar space, and had thought hard about which ones to search for first.

He came to my office at Berkeley, and asked "Should I look for NH_3?" (NH_3 is ammonia gas, the eye watering stuff exuding from bottles of household ammonia.) I had acquired an undeserved reputation for knowledge of the interstellar medium, so I suppose that is why Townes had come. Rising to the challenge, I went to the blackboard and wrote a few equations indicating that to search for NH_3 would be hopeless – there was no way to form enough molecules for him to detect.

Fortunately Townes ignored my advice and, not long afterward, discovered NH_3 in space. Because of its special properties, it has become an important tracer of interstellar clouds which are so dense that they must be on their way to forming stars. The theory for the formation of NH_3 is much more complex than I contemplated in those early days.

Theoretical arguments can't always be trusted in astronomy. The Universe is far more complex than we can imagine in a lifetime.

G.B.F

2
Interstellar Space
Dark Realms of the Nighttime Sky

. . .Beyond the stars of the sixth magnitude you will behold through the telescope a host of other stars, which escape the unassisted sight, so numerous as to be almost beyond belief, for you may see more than six other differences of magnitude, and the largest of these, which I may call stars of the seventh magnitude, or of first magnitude of invisible stars, appear with the aid of the telescope larger and brighter than stars of the second magnitude seen with the unassisted sight. But in order that you may see one or two proofs of the inconceivable manner in which they are crowded together, I have determined to make out a case against two star-clusters, that from them as a specimen you may decide about the rest.

As my first example I had determined to depict the entire constellation of Orion, but I was overwhelmed by the vast quantity of stars and by want of time, and so I have deferred attempting this to another occasion, for there are adjacent to, or scattered among, the old stars more than five hundred new stars within the limits of one or two degrees. For this reason I have selected the three stars in Orion's Belt and the six in his Sword, which have been long well-known groups, and I have added eighty other stars recently discovered in their vicinity, and I have preserved as exactly as possible the intervals between them. . .

As a second example I have depicted the six stars of the constellation Taurus, called the Pleiades (I say six intentionally, since the seventh is scarcely ever visible), a group of stars which is enclosed in the heavens within very narrow precincts. Near these there lie more than forty others invisible to the naked eye, no one of which is much more than half a degree off any of the aforesaid six; of these I have noted only thirty-six in my diagram. I have preserved their intervals, magnitudes, and the distinction between the old and the new stars, just as in the case of the constellation Orion.

From *The Sidereal Messenger,*
by Galileo Galilei, Venice, 1610

When astronomers say "universe", we simply mean this: vast tracts of empty space and enormous stretches of time populated sparsely by stars and galaxies glowing in the dark. The word galaxy is Greek for "Milky Way", the huge group of stars that is home to our Sun and its planetary system. Beyond our own reside other galaxies, each containing hundreds of billions of stars; and although we have so far photographed only the nearest million or so galaxies, our observations of deep space imply that billions of them probably exist.

Where does the Universe end? We don't yet know, but we are fairly sure that we can perceive only so far into deep space. Owing to the finite speed of light, when we look at objects in space, we study them as they were in the past. All evidence indicates that the Universe began at a definite moment some 10 to 20 billion years ago, dubbed the "big bang"; and beyond this time frame, we cannot even hope to probe. This momentous event – this creation – defines the limits of the observable Universe.

The galaxies between here and this observational limit each have their own peculiarities. But like people, they also share many attributes. All seem to be composed of stars, and, among those having a flattened shape, all have a spiral form; among those that are more roundish, almost none have spiral structure. A major problem in contemporary astronomy is how, during the evolution of the Universe since its creation, these "spiral" and "elliptical" galaxies formed, and how they evolved to their present states. Not surprisingly, then, the most reasonable place to begin our study is within our own Galaxy.

Overall Composition

The dark spaces amongst the stars of our nighttime sky are not empty. Throughout our Milky Way Galaxy lurk interstellar clouds harboring vast quantities of gas and dust. These clouds, which have cooled to a temperature of close to absolute zero, are often associated with hot, luminous stars, whose energy outputs are so large that they must have formed only recently in the life of the Milky Way. Astronomers have long thought that they must have emerged from the matter in the clouds where they are found. So the clouds of gas and dust floating in interstellar space are the breeding grounds of the stars described so eloquently by Galileo several centuries ago. Yet the clouds themselves are nearly invisible.

The mechanism of star formation is vitally important to our understanding of the Universe, for nearly all aspects of galaxy evolution hinge upon it. If we can understand how stars originate in the Milky Way, then we should be able to decipher why the youngest stars in particular form a structure having spiral shape and why stellar populations are distinctly different, not only in various parts of our Milky Way Galaxy, but in other galaxies as well. Our Galaxy is the only galaxy for which astronomers can achieve sufficient angular resolution to study star formation in detail, and thus it serves as a prime example of all the universal building blocks we call galaxies.

Sprinkled throughout interstellar space between the dark clouds is a tenuous medium of gas and dust, whose substance is much more rarefied than the dark clouds themselves. This medium, while warmer than the dark clouds, is also exceptionally cold, with temperatures only 100 degrees or so above absolute zero.

Here and there, where hot stars happen to encounter it, this medium is heated to many thousands of degrees, thus forming the great glowing regions known as emission nebulae. These are the spectacular and colorful objects often adorning the centerfolds of astronomy textbooks. Heated in this way, the interstellar gas forms a plasma of electrically charged ions and electrons, which radiate emission features at specific wavelengths characteristic of whatever chemical elements are present. Analyzing such radiation, astronomers have found that interstellar gas is mostly made of hydrogen and helium, along with traces of heavier elements such as carbon, nitrogen and oxygen. This is hardly surprising, since our Sun and most other stars have very similar compositions. Taken together, all the data support the idea that stars form from interstellar matter.

As their name implies, emission nebulae are visible, and thus they have been studied for nearly a century by optical astronomers. However, at any given time, most interstellar gas is too remote from any hot stars to be heated enough to shine visibly, and so the gas must be studied by other methods. A powerful technique first used in 1951 employs the faint emission feature of hydrogen atoms (H) at a radio wavelength of 21 centimeters (about the length of this book) to map the general spread of gas in our Galaxy on a grand scale. The results delineate the great spiral arms that emanate from a central region and wind outward over a distance of some 100,000 light-years; the arms coincide with those outlined by dark clouds and young stars, and show that there is enough atomic hydrogen in space to form 5 billion suns. Thus the pioneering radio astronomers of the 1950s revealed a world completely invisible to Galileo's telescope, and indeed, to every optical telescope built since that time; their observations further support the notion that stars form from gas and dust in interstellar space.

In the 1960s, 21-centimeter observers discovered to their surprise that individual dark clouds often contain little atomic hydrogen. Why should this be? After all, observations reveal that such clouds are richly endowed with particles of interstellar dust that obscure the stars behind them. From the way that dust obscuration depends upon wavelength, we can infer that the size of typical dust particles ranges between 10^{-6} and 10^{-5} centimeter – somewhat smaller than the 10^{-5} to 10^{-3} centimeter size of soot or house dust. Because interstellar dust is thought to be made of silicon, iron, magnesium, and oxygen (all of which have much lower abundances in the Universe than hydrogen), hydrogen gas is expected to accompany such dust in space. Theorists had earlier suggested that many of the hydrogen atoms react together on the surfaces of dust particles to form hydrogen molecules (H_2); this type of hydrogen does not emit 21-centimeter radiation, and thus would escape detection by radio astronomers. Suspicions were heightened when, during the 1970s, radio observers discovered molecules such as carbon monoxide (CO) in dark interstellar clouds. The CO data proved that the total amount of gas in dark clouds must be much greater than that derived from the 21-centimeter observations of neutral hydrogen. Given the preponderance of hydrogen in the Universe, it seemed safe to assume that molecular hydrogen is the dominant component of the dark interstellar clouds.

In 1973, an ultraviolet telescope aboard NASA's *Orbiting Astronomical Observatory* # 3 (renamed *Copernicus* in honor of the great astronomer born 500 years earlier) confirmed the finding by an earlier rocket flight that H_2 is present in interstellar space, and proved for the first time that H_2 is in fact widespread throughout our Galaxy. The satellite did so by detecting narrow absorption features characteristic of H_2 as radiation from distant hot stars passes through interstellar clouds between us and the stars. (An absorption feature or "line" is a narrow spectral interval in which radiation is absorbed while passing through matter; the wavelength of the feature is uniquely related to the composition of the matter). The abundance of H_2 was found to be correlated with that of dust, confirming the theoretical prediction that hydrogen atoms react on the surfaces of dust grains to become molecular hydrogen. From the ultraviolet data, we calculate that there is enough H_2 to form another 5 billion stars having the mass of the Sun. Counting both atoms and molecules, then, some 10 billion solar masses of interstellar hydrogen populate our Galaxy.

Thus our understanding of interstellar matter has changed dramatically as a result of radio and ultraviolet observations in recent years. Whereas astronomers had previously regarded the emission nebulae as the largest individual entities of interstellar gas, we now know that, despite their huge extent, such nebulae are mere islands of hot gas within an ocean of cool gas composed of roughly equal amounts of atomic and molecular hydrogen.

Observations by *Copernicus* also determined the abundances of many other gases in interstellar space. Again using its ultraviolet telescope, this satellite proved that the abundances of many chemical elements - at least the amounts in gaseous form - are much lower than in our Solar System and in stars. The most likely explanation for this unexpected finding is that substantial quantities of familiar elements such as carbon, oxygen, silicon, magnesium, and iron have been used up in forming grains of interstellar dust. This idea is supported by observations of objects at wavelengths near 10 microns in the infrared region, which reveal strong absorption features characteristic of iron and magnesium silicates. If these infrared features are indeed caused by interstellar dust, they account for at least a fraction of the iron, magnesium, silicon, and oxygen which is missing from interstellar gas. Even so, the complete chemical identification of interstellar dust remains an unsolved problem.

Interstellar Molecules

Throughout the 1970s, radio astronomers discovered a pharmaceutical array of molecules in the dark clouds of our Milky Way. The molecules are detected primarily as they emit or absorb radiation while changing their energy of rotation in space. Like most other physical systems from quarks to quasars, molecules must change their states to be detectable.

Now totaling more than 60 species, the list of interstellar molecules includes familiar substances such as ammonia (NH_3) and water vapor (H_2O). These chemicals are thought to have played important roles in the prebiological chemical processes that led to life on Earth, so their presence in space is intriguing. Even more suggestive is the presence of a variety of organic molecules that (like those of living organisms) contain carbon. The most complex organic molecule observed in interstellar space to date is cyano-deca-penta-yne ($HC_{11}N$), which although small by biological standards, has a size which theorists have difficulty explaining in an astronomical context. No organic ring molecules have yet been discovered.

Thus far, nearly all the molecules have been discovered in dense, dark, and dusty clouds composed mainly of H_2, so astronomers now refer to such regions as molecular clouds. The dust absorbs and scatters short-wavelength radiation, including light, making the dark clouds inaccessible to optical astronomy. But almost all the molecules emit radio waves that can escape the dusty regions without attenuation. Hence, the radio study of molecules gives us a powerful method to probe these rather mysterious regions in space.

The simple molecule carbon monoxide (CO) has been found to be distributed nearly ubiquitously in our Galaxy. It is especially abundant within clouds of the spiral arms, and so it has been used to map those arms. The CO results agree well with those of the 21-centimeter H observations noted above. In addition, radio maps of CO (usually made by observing at wavelengths near 2.6 millimeters) reveal that molecular clouds do not reside alone in space, but in huge complexes, some spanning as much as 150 light-years across and harboring roughly a million solar masses of gas. These complexes, which number about a thousand in our Galaxy, are usually elongated parallel to the plane of our Galaxy, each complex containing numerous individual clouds some 30 to 60 light-years across. In turn, within these individual clouds lurk much more compressed – and extremely dark – regions often less than a light-year in diameter. It is in these compact "cloud cores" where star formation is now observed to be occurring.

Because long-wavelength radiation can penetrate cloud cores, radio and infrared astronomers have made most of the discoveries regarding the earliest stages of star formation, a subject about which optical astronomers could only speculate. Infrared emission is observed from objects which cannot be identified with any optical objects, but which also emit extremely intense radio spectral features of hydroxyl (OH) and water vapor (H_2O) molecules. The energy sources for these infrared objects seem to be optically luminous hot stars which, however, are hidden from view by nearby dark clouds. Apparently what is happening is that the stars are so hot that they emit large amounts of ultraviolet radiation which is largely absorbed by a surrounding "cocoon" of dust. The absorbed energy is then reemitted by the dust as infrared radiation. Some of the ultraviolet radiation heats and ionizes the accompanying hydrogen, whose emission can be observed in the radio domain. The OH and H_2O spectral features result from the action of a natural maser – the radio equivalent of a laser – enabling the clouds to cool off in spite of the heating by the stars. Some of the clouds are so massive that their own gravitation is trying to collapse them further. As this tendency is resisted mainly by the random motions of the molecules, it is possible that the cooling provided by the escape of maser radiation plays a significant role in the collapse process. That dust cocoons are invariably found in the dense cores of molecular clouds supports the idea that the hot stars responsible for their heating only recently emerged from the surrounding cloud.

Do not lose sight of the fact that all this happens in the darkest regions of galactic gas and dust. Not unlike the mammals, the brightest stars are incubated in total darkness.

Molecular Clouds and Star Formation

We are still uncertain how most interstellar molecules form. The subject of astrochemistry, with its unfamiliar nonequilibrium conditions and unearthly temperatures and densities, is only now being developed for the first time. Those books on our shelves labeled "General Chemistry" are mistitled; they should really be called "Earth Chemistry", as we are only now beginning to deal with a truly general or universal chemistry. As a result of our studies of interstellar molecules, astronomers are the ones posing the problems for this new chemistry.

Apparently interstellar molecule formation begins when hydrogen atoms combine to form H_2 on the surface of dust grains. From time to time, some of these hydrogen molecules are hit by charged particles (known as cosmic rays) moving at nearly the speed of light. Such particles were discovered in the vicinity of the Earth many years ago, though we now presume that they pervade the entire Galaxy. Cosmic rays are made, like stars, largely of hydrogen and helium from which electrons have been stripped to yield protons and helium nuclei. When such a particle hits an interstellar H_2 molecule, an electron is knocked off, forming a hydrogen molecular ion, H_2^+. The H_2^+ then attaches to another H_2, forming H_3^+ and a hydrogen atom; the H_3^+ reacts in turn with carbon (C), nitrogen (N) and oxygen (O) to produce species such as the formyl ion (HCO^+), hydrogen cyanide (HCN), and formaldehyde (H_2CO). Further reactions result in CH_4, NH_3, H_2O, *etc.* Complete chains of reactions are understood for the simplest molecules, but the heaviest molecules still challenge our theoretical understanding. Conceivably, some of the more complex interstellar molecules are actually fragments of even larger molecules not yet discovered, but how the huge molecules would form is a mystery.

Because many of the interstellar molecules are good radiators, their abundance determines the rate at which molecular clouds cool down; this in turn determines whether the clouds will collapse under their own weight, because the random motions of molecules that occur at higher temperatures oppose collapse. A key theoretical problem is the balance among the various forces operating in molecular cloud cores. Observations of CO and other molecules permit astronomers to estimate the total amount of matter present, and hence the magnitude of the inward-directed gravitational forces within a cloud core. The temperature can also be derived from CO observations, yielding estimates of the random motions of the molecules. When all these data are compared, we find that, for most cloud cores, the inward gravitational forces are strong enough to overcome the random motions, so that most cores should now be collapsing. The collapse can be calculated to take only a few million years. If we are correct in arguing that the collapse of molecular cloud cores results in star formation, we find that the number of young stars in the Galaxy would be almost a hundred times larger than observed. Accordingly, our argument must somehow be wrong.

One way out of this contradiction invokes the concept of turbulence; although the random motions of individual molecules comprising cloud cores are insufficient to prevent them from collapsing into stars, large-scale motions like those typifying the raging rapids of a swiftly flowing river – which we call turbulence – might be able to resist collapse, and thereby keep cloud cores in a state of suspended equilibrium; the process might be akin to the way that a juggler keeps several balls in the air, thus seemingly defying gravity as long as the balls are hurled upward. That turbulence is present in the cloud cores is well known, because radio astronomers can use the wavelength shifts in the observed molecular features to derive the random velocities; the results are always larger than the random motions due to temperature alone. For some clouds, turbulence in fact seems adequate to resist gravitational collapse.

For this explanation to be valid, the turbulent velocities must be much higher than the speed of sound. Just as in the case of supersonic aircraft in the atmosphere of Earth, turbulent motions of such magnitude should set up shock waves which theorists predict would rapidly dissipate the turbulent motions, bringing us back to the original contradiction between the inferred high rate of collapse and the observed low rate of star formation. Possibly, the effects of shock waves could be much reduced if molecular clouds are actually composed of huge numbers of "cloudlets" too small to resolve with present instruments. In this model, the cloudlets would whiz by each other without hitting, like balls from the juggler's hand, thus minimizing the shock waves. Radio astronomers would dearly like to have new radio telescopes able to resolve any individual cloudlets.

A currently popular theory of star formation stipulates that cores form within clouds when nearby hot stars of a recently formed generation of stars ionizes and heats that part of the parent molecular cloud nearest to them. This drives a shock wave into the cloud, compressing the gas and raising its local density above a critical value needed for gravity to become important and thus start the cloud collapsing. The gas then contracts to form a cloud core, part of which collapses to trigger a new generation of massive stars, and the cycle is repeated. In this way, a wave of successive star-forming events propagates through a complex of molecular clouds. This idea is supported by the fact that alignments of stars are observed near the outside rims of various molecular clouds; groups of stars nearest such clouds appear to be the youngest, while those farther away appear to be older, pretty much like what is expected.

Hot Bubbles

Besides determining the chemical composition of the interstellar gas, the *Copernicus* satellite discovered a small amount of gas that is far hotter than anyone expected. While using *Copernicus* to observe ultraviolet radiation from distant hot stars, astronomers noticed some absorption features characteristic of oxygen atoms having five of their normal complement of eight electrons stripped away. Such highly ionized atoms could not reside in ordinary interstellar clouds, where the temperatures are only about 100 degrees, or even in emission nebulae, where the temperatures reach ten thousand degrees or so. To remove five electrons from oxygen, the temperature must be on the order of 500,000 degrees. Apparently, hot ionized gas of this temperature forms an "intercloud medium" between the clouds of atomic and molecular hydrogen. Extending far into space beyond our local neighborhood, this thin but super-heated gas might conceivably stretch even into the vast intergalactic space between the galaxies.

Closer to home, x-ray observations have also revealed a hot gas permeating the local interstellar space just beyond the farthest known planet, Pluto. There, much of the rarefied interstellar gas is hot enough to emit x rays – at about a million degrees temperature – suggesting that our Solar System is virtually engulfed by a sea of seething gas.

Without our newly invented techniques to detect invisible radiation, we could not have imagined, even in our wildest dreams, that our Solar System resides inside a bubble of hot gas. But as we shall see in Chapter 4, this conclusion is in harmony with a completely different line of investigation suggesting that our Solar System may well have originated when the concussion of an exploding star, called a supernova, triggered the collapse of a molecular cloud some five billion years ago. Supernova explosions are just what are needed to heat the intercloud gas to hundreds of thousands or even a million degrees. In fact, we have evidence that supernovae send shock waves into the surrounding medium, heating some of it to enormous temperatures, and occasionally triggering the collapse of molecular clouds. This evidence comes from radio, optical, and x-ray studies of the remains of supernovae. Radio astronomers were the first to detect radiation from such regions – radiation that arises from fast particles accelerated in the stellar explosion. X-ray astronomers pick up the radiation from the very hot gas heated by the shock wave propagating into the surrounding interstellar gas, while optical astronomers see the radiation from the same gas after it has cooled a bit.

Our current understanding of matter in interstellar space is based, like many aspects of science, on considerations of energy conservation. Energy that is lost by radiation from interstellar matter must be compensated somehow from other sources. What sources of energy can we identify for heating the interstellar medium? We have already spoken of the ultraviolet radiation emitted by hot stars, the ubiquitous cosmic-ray particles that regularly collide with the atoms and molecules of the gas, and the rapidly expanding shock waves caused by the explosive deaths of previously bright stars. When these sources are included in a theoretical model, we find that the diffuse clouds of atomic hydrogen can be maintained at about 100 degrees above absolute zero, as observed, by ultraviolet starlight which penetrates them, and which is able to eject energetic electrons from dust grains, thus heating the gas. Molecular clouds are much colder, typically only 20 degrees above absolute zero, because ultraviolet photons cannot penetrate the large amounts of dust present, leaving cosmic-ray particles as the only adequate source of heat. Emission nebulae reach equilibrium at much higher temperatures – about 10,000 degrees – because they have enough ultraviolet radiation from their embedded hot stars to eject electrons from all the hydrogen atoms present, thus heating the gas very effectively. And the intercloud medium is maintained at a temperature of 500,000 degrees or more by shock waves propagating from exploding stars.

Thus, we have rather recently come to regard the interstellar medium as composed of several quite different phases, each having widely differing temperatures and densities. Perhaps the closest analogy to this situation we encounter on Earth are the different phases in which we find a substance like water. At very low temperatures, water forms a solid – ice –, at intermediate temperatures it is liquid, and above the boiling point it becomes a gas – steam. While currently trying to develop consistent multiphase models of the interstellar medium, theorists are grappling with an array of questions. What happens at the boundary between two different phases of interstellar gas? How do the different phases exchange mass and energy? What determines the proportion of matter involved in each of the observed phases? Beyond these, several other more general questions confront the interstellar theorist. What mechanisms stabilize molecular clouds? Are supernova shock waves required to trigger gravitational collapse? Why do the molecular cloud cores fragment while contracting into objects whose masses are those of stars? What determines the final distribution of stellar masses? Do some fragments other than the one that led to our Solar System become planetary systems, while still others become binary- or multiple-star systems or even isolated stars all without planets? Telescopes of the future – especially those able to detect invisible radiation such as x rays from hot interstellar gas, ultraviolet spectral features in all phases of the interstellar medium, and radio and infrared emission from molecular clouds – will greatly increase our understanding of these most basic questions.

Prospects for the Future

Placement into orbit of the *Space Telescope* by NASA's *Space Shuttle* in the mid-1980s should provide us with an unparalleled platform from which to conduct ultraviolet observations of gas in interstellar space. As detailed in Appendix A, the greater aperture of the *Space Telescope*, together with the ability of its detectors to record information simultaneously at many wavelengths, insures that its sensitivity will be much greater than that of *Copernicus*, enabling us to study interstellar matter along lines of sight both to more distant and to more heavily obscured stars. The wider spectral coverage of a smaller telescope embodying a far-ultraviolet spectrograph, which is proposed to accompany *Space Telescope* into orbit, should also permit us to measure the spectral features of large numbers of different atoms, ions, and molecules in interstellar space – species that have important bearing upon the state and motions of interstellar clouds, and upon theoretical models for how molecules form within those clouds.

To study molecules that radiate at short radio wavelengths, astronomers intend to open a new part of the electromagnetic spectrum – the submillimeter wavelength domain. An instrument operating in this domain will permit observation of several key interstellar molecules that radiate at wavelengths in selected atmospheric transmission bands down to 0.3 millimeters. To achieve sensitivity sufficient to allow meaningful measurements and to permit maps to be made with an angular resolution of at least 10 arc seconds, the *Submillimeter-Wave Radio Telescope* described in Appendix B will have to be roughly 10 meters in diameter.

This proposed radio telescope would allow us to attack several problems noted earlier in this chapter. For example, it would resolve distances as small as 0.1 light-year in the active star-forming clouds of Orion. This could well make it possible to detect any cloudlets which theorists think comprise molecular clouds; in turn this would perhaps enable us to answer the thorny question of what supports such clouds against gravitational collapse. The added sensitivity and increased wavelength coverage of this telescope are likely to provide important tests of the chains of chemical reactions that have been theoretically predicted to give rise to the many types of molecules observed in interstellar space; only by observing spectral features of a number of different molecular species whose abundance is predicted to depend in different ways on local conditions can we hope to push forward the newly emerging subject of astrochemistry. Furthermore, because of its greatly increased angular resolution, such a submillimeter radio telescope will be invaluable in clarifying the spatial relationships among molecular cloud cores, emission nebulae, and young stars in star-forming regions. For studies beyond the Milky Way, it will extend the frontier of the young subject of extragalactic molecular astronomy, and especially our efforts to map the distribution of such clouds in other galaxies.

Even with these two major instruments, we would still not be well suited to study the most abundant molecule of all – H_2. The far-ultraviolet spectrograph, in accompaniment with the *Space Telescope,* will be able to extend our knowledge of H_2 in space, but only in those directions where bright hot stars happen to be located along our line of sight; in any other direction, there are no sources of suitably intense ultraviolet radiation capable of being absorbed by interstellar H_2. On a more positive note, the fundamental emission feature of H_2 occurs at a wavelength of 2.8×10^{-3} centimeter, deep in the infrared. Soon, infrared instruments carried aloft by aircraft and balloons will attempt to make the first measurements of this basic spectral feature, which until now has been unobservable because our atmosphere absorbs so heavily at this wavelength. An infrared telescope in space, cooled almost to absolute zero to eliminate unwanted (heat) radiation from the telescope itself, and equipped with an array of detectors tuned to the fundamental feature of H_2, should permit us to map interstellar H_2 over the entire Galaxy much as H has been mapped in the radio domain at 21 centimeters wavelength and CO at 2.6 millimeters wavelength. Since H_2 is the primary component of star-forming clouds, we should then be able to observe star formation throughout the whole Galaxy, and not just in those few regions that happen to lie near us. An infrared instrument such as described is now being readied; the vital statistics of this *Shuttle Infrared Telescope Facility* are listed in Appendix C.

Of considerable importance in our quest to decipher the intricacies of star formation, this orbiting infrared laboratory will surely search for the infrared emission from cocoons around newly-formed stars in the cores of molecular clouds. With its thousand-fold better sensitivity than any infrared equipment currently available, the *Shuttle Infrared Telescope Facility* should be able to extend the search to much fainter limits than are now possible. Furthermore, we should be able to address directly the issue of cloud fragmentation, making a special effort to decipher the range of stellar masses that emerge from a cloud of given mass. And the enormous ability of this infrared facility to sort out different wavelengths with great precision should enable astronomers to detect and study a great variety of molecules, some of which are expected to radiate effectively only when the gas density is rather high, much like that expected in clouds already collapsing to form stars. Thus the *Shuttle Infrared Telescope Facility* will be used to spot nascent stars before they emerge from the relatively cool gaseous regions of our Galaxy.

As for observations of the very hot gas in interstellar space, the *Advanced X-ray Astrophysics Facility* described in Appendix D, now being designed for a late-1980s launch, will permit observations of individual x-ray emission features in shock waves around supernovae – the kind of measurements that x-ray astronomers have been denied to date because both the size and lifetime of previous x-ray instruments have been insufficient. Such observations should allow us to better understand the processes occuring behind such shock waves. One such process might be the destruction of interstellar dust grains – which are responsible for the obscuration or light in dark clouds – by the surrounding, shock-heated gas. This process might explain why not all the atoms of chemical elements like carbon and oxygen are locked up in such grains; occasionally the atoms are knocked off grains by passing supernova shock waves, and only slowly attach to grains again. Of considerable interest to residents of the Solar System, the *Advanced X-ray Astrophysics Facility* will enable us to map and thus better understand the superhot gas that engulfs us all.

Despite their apparent diversity, these studies will continue to address a unifying theme: the varying interrelationships of the many components of interstellar matter. Comprising nothing less than a "galactic ecosystem", the evolutionary balance between these components might be as complex and delicate as that of life in a tidepool or a tropical forest. Only by being receptive to information from the Milky Way in all the electromagnetic wavelength bands in which it chooses to radiate can we hope to understand some of nature's cycles within our Galaxy: The flow of gas from dense interstellar clouds into newborn stars and its return to interstellar space by means of supernova explosions; the chemical cycles establishing the abundances of interstellar molecules; and the flow of energy from hot and sometimes exploding stars, which inject enough energy to keep the interstellar gas at reasonable temperatures. These are but a few of the issues that will drive astronomical research during the remaining years of this century.

Whatever progress is made, we can be virtually certain that the instruments constructed so as to address these issues will also discover completely unsuspected phenomena in space. From the time of Galileo, surprise has been among the hallmarks of astronomy.

Collisions of Interstellar Clouds

The great astronomer Jan Oort of the Leiden Observatory in the Netherlands had written a paper sketching a theory of star formation. According to his theory, clouds of interstellar gas float in space, being held together by the pressure of a thin but hot intercloud medium. From time to time two clouds collide, dissipate their energy, and merge, so that statistically there is a tendency for clouds to become ever more massive. When a given cloud becomes massive enough, the mutual gravitation among its parts causes it to collapse, forming some stars. The most massive among the stars are very luminous, and send shock waves into the neighboring gas, breaking it up into a new generation of small clouds; the cycle is then repeated. Such a model would account for the long-term persistence of both interstellar clouds and star formation.

Thinking about this model as a young professor at Princeton, I saw a way to cast it into mathematical terms. However, the resulting equations appeared too difficult to solve, so I requested the assistance of a very capable undergraduate, William Saslaw. He immediately programmed the problem on a computer, and came in with the solution. Saslaw's solution seemed to be surprisingly simple, so I kept wondering if it could be obtained without a computer. Finally I found an exact solution using a mathematical technique I had learned as an undergraduate at M.I.T., but which I had not used since. The paper I wrote with Saslaw used my solutions, illustrated by his computations.

Much later, my friend and mentor Lyman Spitzer reviewed this area of research in a book he was writing on interstellar matter. He was successful in deriving an approximate form of the solution from an elementary physical argument.

Astronomers use computers a lot in their work, and they yield results which can often be obtained in no other way. Still, solutions based on analytical reasoning and physical insight will continue to have much to teach us also.

G.B.F.

3
Sun and Stars
Sites of Invisible Activity

Sunspots are generated and decay in longer and shorter periods; some condense and others greatly expand from day to day; they change their shapes, and some of these are most irregular; here their obscurity is greater and there less. They must be simply enormous in bulk, being either on the sun or very close to it. By their uneven opacity they are capable of impeding the sunlight in differing degrees; and sometimes many spots are produced, sometimes few, sometimes none at all.

Now of all the things found with us, only clouds are vast and immense, are produced and dissolved in brief times, endure for long or short periods, expand and contract, easily change shape, and are more dense and opaque in some places and less so in others. Indeed, all other materials not only lack these properties but are far from having them. Moreover there is no doubt that if the earth shone with its own light and not by that of the sun, then to anyone who looked at it from afar it would exhibit congruent appearances. For as now this country and now that was covered by clouds, it would appear to be strewn with dark spots that would impede the terrestrial splendor more or less according to the greater or less density of their parts. These spots would be seen darker here and less dark there, now more numerous and again less so, now spread out and now restricted; and if the earth revolved upon an axis, they would follow its motion. And since clouds are of no great depth with respect to the breadth in which they are normally extended, those seen at the center of the visible hemisphere would appear quite broad, while those toward the edges would look narrower. In a word, no phenomena would be perceived that are not likewise seen in sunspots.

From *History and Demonstrations Concerning Sunspots and Their Phenomena*
by Galileo Galilei, Rome, 1613

STUDIES OF THE SUN differ in two major ways from the other disciplines of astronomy. First, because of the enormous quantity of radiation we receive from our nearest star, the scientific program of solar physics can readily utilize the full range of observational techniques available to astronomers. As a result, solar observations now cover the entire electromagnetic spectrum, from long-wavelength radio waves generated by electrons moving in the Sun's outer atmosphere, to ultra short-wavelength gamma-ray emissions arising from collisions between atomic nuclei in solar flares. Measurements are made routinely not only with telescopes on Earth's surface but also by satellites in Earth orbit. Other spacecraft venture into the Solar System itself, orbiting the Sun like little planets, to sample streams of solar gas in space.

Second, unlike most other branches of astronomy, solar physics is closely connected to practical concerns, such as the Earth and its climate, as solar energy and material interact directly with Earth. Hence, activity on the Sun, which is largely invisible, is important to predict if at all possible.

Solar Activity

For a century, scientists have known of gigantic eruptions from the surface of the Sun. Somewhat more recently, it became clear that well-known terrestrial phenomena, including the Aurora Borealis and radio blackouts, owe their origin to such solar storms. As a result of space research, we now realize that such eruptions release vast quantities of matter and energy, flooding the planetary system with fast particles and invisible radiation. Indeed, the Sun is the site of intensely energetic and extraordinarily complex activity, much of which cannot be readily seen by optical means.

The Sun's outermost atmosphere, called the solar corona, is normally invisible because the bright disk of the Sun *per se,* scattered in our atmosphere, overwhelms the much fainter corona with its glare. Even so, during total eclipses of the Sun, when the Moon conveniently blocks our view of the solar disk itself, the corona becomes startlingly obvious against the now-dark sky. Since the spectrum of sunlight changes abruptly at the moment an eclipse becomes total, thus revealing the existence of highly ionized gases in the corona, we have also come to realize that the solar corona has an astonishingly high temperature; it is at least a million degrees. One of the central problems in solar physics is this: If the Sun's surface temperature is "only" 6000 degrees, how does its outermost atmosphere maintain an enormously higher temperature? After all, temperatures normally drop off as one moves away from a source of heat – and our Sun is surely just that. The culprit, we think, is the partly regular and partly sporadic activity at the surface (called the photosphere) of the Sun. To the naked eye, or even a small telescope, the solar surface looks smooth and uneventful. Bright, yes, and hot enough to vaporize steel, but calm. With better instruments we have learned that this appearance is deceptive. Small motions of gases at the surface have become giant crashing waves when they reach the upper levels of the solar atmosphere. And sunspots – those dark regions first studied by Galileo – are just hints of a curious type of engine operating deep in the Sun, which produces strong magnetic fields at its surface. Such phenomena – which vary with time – are collectively called "solar activity".

As instruments have improved over the past few decades, many varied aspects of solar activity have been discovered. Besides eruptions and the long-studied sunspots, we gradually became acquainted with quiescent prominences, solar flares, plages, and other kinds of disturbances that delineate the active regions on our Sun. Observations showed us that the sunspots are governed by strong magnetic fields, and that flares are the source of highly energetic particles. The Sun's general magnetic field was mapped, found to vary periodically every 22 years, and monitored throughout an entire 22-year cycle. And, eventually, the intimate connection of solar magnetic fields with solar activity – super-heated gases, fast particles, violent eruptions, and spectacular flares – has become qualitatively clear. Quantitatively, the problem of solar activity is still unsolved, largely because it is so complex. Perhaps terrestrial weather is an appropriate analogy. We know the phenomena – clouds, winds, thunderstorms, tornados – but our understanding is a long way from being able to predict "when and where" such phenomena will occur.

Curiously enough, completely unrelated objects – comets – provided the first hint that the solar corona is expanding into space. Comet tails always point away from the Sun, even on their outbound journeys, suggesting that something must push the spent cometary debris away from the Sun. In many cases, we can calculate that an adequate push is provided by solar radiation, but in others the radiation is clearly inadequate, so it was postulated that gas streams away from the Sun, forming a "wind" not unlike the breezes we know so well on Earth. In a key accomplishment of the Space Age, Earth-orbiting spacecraft have directly measured the solar wind that does the pushing. From these measurements we infer that the solar wind carries away a million tons of solar material every second at velocities that vary, but which average 500 kilometers per second. As far as we can tell, the composition of the solar wind is the same as the makeup of the Sun itself, inferred from spectra of its surface – largely hydrogen and helium, with traces of heavier elements.

Does the solar wind mean that our Sun will not last very long? Not at all, for the Sun, weighing in at 300,000 Earths, has a mass far greater than anything with which we are familiar. Thus, even at this prodigious rate, the solar wind will carry away less than one-tenth of one percent of the total bulk of the Sun throughout its expected lifetime of ten billion years.

The solar wind extends far into space, forming a so-called "heliosphere" within the surrounding interstellar medium – a sea of plasma some hundred times the size of Earth's orbit. Filled with streams of coronal gas that crash into one another violently to produce shock waves and energetic particles, the heliosphere is constantly energized by blasts from solar flares. When the solar wind interacts with the (by comparison nearly insignificant) planets, some of the charged particles in the wind become trapped in the outskirts of planetary magnetic fields, thereby producing huge zones of charged particles girdling the midsections of many planets. The Van Allen Belts that hover invisibly above the Earth are a good example. A few days after especially intense solar flares, these belts often become overloaded with solar particles, thus generating severe movements in Earth's magnetic field. Unbeknownst to us on the surface because none of our senses respond to magnetic fields, such "magnetic storms" can nonetheless wreak havoc with electrical transmission systems. Occasionally, energetic particles leak out of the swollen belts and crash into the upper atmosphere of the Earth, causing spectacular aurorae to flash across the nighttime sky. The direct radiation from solar flares, in the ultraviolet and x-ray bands, can severely affect the layers of our atmosphere called the ionosphere which we use to reflect radio transmissions around the globe. No wonder that government agencies concerned with radio communications take an interest in solar activity.

Despite the regularity of the 22-year solar cycle (which is manifested in an 11-year cycle in the number of sunspots), much of the Sun's truly violent activity is irregular. For example, in the mid-1970s, instruments aboard the *Skylab* orbiting space station revealed that the solar corona erupts every few hours. X-ray observations made from this manned satellite outlined in dramatic detail the hot, dense knots of the active corona and especially their association with the solar magnetic field. Perhaps more important, *Skylab* also discovered abnormally cool, thin regions in the corona, called coronal holes, from which the solar wind of particularly high speed streams forth. Here we have an example of our growing ability to predict solar activity. When coronal holes are detected by our telescopes, we can be fairly sure that a high-speed solar wind stream will encounter the planets in the near future.

Instruments aboard NASA's *Solar Maximum Mission* satellite have just completed a series of precise, coordinated x-ray and ultraviolet observations of solar flares. As a result, we know that such short-wavelength radiations are especially intense when emanating from the extremely compact hearts of solar flares, where temperatures can reach a hundred million degrees. These and other observations of invisible plasma now aid our understanding of flare eruptions on our Sun. They seem to be explosions that not only replenish the corona constantly but also have substantial impact on the whole Solar System.

Another recent discovery is that the level of our Sun's activity also seems to slowly rise and fall over centuries. Historical studies have shown that solar activity reached an abnormally low level during the fifteenth and seventeenth centuries, while rising to a very high level during the twelfth century. What's more, the climate of Earth seems to respond directly to these long-term solar variations, with Earth's mean annual temperature declining during extended periods of low solar activity. The causes of such century-long solar variations remain a mystery; moreover, it is unclear through what mechanism they affect terrestrial climate.

What is the origin of the magnetic fields inside the Sun? What roles do they play in flares, prominences, sunspots, and other features of our active Sun? How do the fields control and perhaps heat the corona? And why does the corona expand to produce the solar wind? Astronomers have made much progress over the past few decades in answering some aspects of these questions, but others evade our understanding to this day.

Stellar Activity and Mass Loss

Our Sun's activity is apparently not unique, for stars of all types are now being found to be active and to have stellar winds. The importance and ubiquity of strong stellar winds became apparent only through advances in spaceborne ultraviolet and x-ray astronomy as well as radio and infrared ground-based astronomy. For example, in the late 1970s and early 80s, the *Einstein* orbiting observatory quite unexpectedly detected x rays from nearly all types of stars, thus revealing that that they too are surrounded by coronas having temperatures of a million degrees or more. Though it may be hard to appreciate while gazing at seemingly quiescent stars in the nighttime sky, apparently all stars sport active regions, including spots, flares and prominences much like those of our Sun. As in the case of our Sun, such activity is largely invisible. In retrospect, we should be hardly surprised, for, by now, the doctrine that our Sun is an ordinary, average, highly typical star has become firmly entrenched in scientific thought. Still, it is exciting to see that philosophically reasonable viewpoint vindicated by direct observation. Some stars exhibit "starspots" so large that an entire face of the star is relatively dark, while others display flare activity thousands of times more intense than that on our Sun.

Furthermore, ultraviolet observations made by the *Copernicus* satellite showed key stellar absorption features to be shifted in wavelength by the

Doppler Effect*, thus demonstrating that many stars have stellar winds, much like our Sun; in some cases, particularly for the most luminous stars, the wind is strong enough to have profound effects, not only on the stars' immediate environment including any attendant planets, but also on the subsequent evolution of the stars themselves. Even stellar winds as weak as the Sun's are able to carry away significant amounts of angular momentum (which is the tendency of material in a rotating body to continue rotating). Enough angular momentum is carried off in the solar wind that the rate at which the Sun is spinning must be slowed over the course of a few billion years. Winds much more powerful than the Sun's can carry away so much angular momentum that the stellar spin is slowed in only a few million years. Such winds can also carry away a substantial portion of the matter inside the star, causing the star to evolve quite differently than its cousins that do not have strong stellar winds.

We now realize that, among the luminous stars, both hot, blue stars and cool, red stars have stellar winds. Consider for a moment the highly luminous, hot, blue stars that have by far the strongest stellar winds. Observations of their ultraviolet spectra with telescopes on rockets and satellites have shown their wind speeds often reach 3000 kilometers per second, while losing mass at rates up to a billion times that of our Sun's wind. Such blasts are of gale force, even when we speak of weather in the Galaxy. The corresponding mass-loss rates approach and sometimes exceed 10^{-5} solar masses per year, one entire solar mass (perhaps a tenth of the total mass in the star) being carried into to space in the – astronomically speaking – short span of a hundred thousand years. Accordingly, we expect the most luminous stars to lose substantial fractions of their mass during their lifetimes, as the latter are calculated to be only a few million years.

Observations made by the *International Ultraviolet Explorer* satellite, a multinational spacecraft currently orbiting Earth, have shown that to produce such great winds the pressure of hot gases in a corona, which drives the solar wind, does not suffice. Instead, the winds of the luminous hot stars must be driven directly by the pressure of the energetic ultraviolet radiation emitted by these stars. The same mechanism has been theorized to eject gas from the cores of active galaxies and quasars, a subject to which we shall return in Chapter 6.

Such powerful stellar winds hollow out vast cavities in the interstellar gas, pushing outward expanding shells of galactic matter resembling those generated by supernova explosions. Aside from the simple observa-

* If the absorbing or emitting gas is moving toward (or away from) the observer, all spectral features are shifted to shorter (or longer) wavelengths; this "Doppler Effect" has been verified accurately in the laboratory.

tion that copious quantities of ultraviolet radiation are available from luminous hot stars to drive stellar winds, the details are not well understood. Whatever is going on, it is surely complex, for the ultraviolet spectra of the stars tend to vary with time, implying that the wind is not steady – much like storms on Earth. In an effort to better understand the variations in the rate of flow, theorists are now investigating possible kinds of instabilities that might be peculiar to luminous hot stars.

Observations made with radio and infrared as well as optical telescopes prove that luminous cool stars also have winds whose total mass flow rates are comparable to those of the luminous hot stars, although their velocities are much lower, about 30 kilometers per second. Because luminous red stars are inherently cool objects (about 3000 degrees surface temperature, or half that of our Sun), they emit no detectable ultraviolet or x-ray radiation, so the mechanism driving the winds must differ from that in luminous hot stars. Unlike the hot stars, winds from luminous cool stars are rich in dust grains and molecules; since nearly all stars more massive than the Sun eventually evolve into such stars, these winds, pouring into space from vast numbers of stars, provide a major source of new gas and dust in interstellar space, thus furnishing a vital link in the cycle of star formation and galactic evolution. As in the case of the hot stars, the specific mechanism that drives the winds of the cool stars is not understood; at this time, we can only surmise that gas turbulence and/or magnetic fields in the atmospheres of these stars are somehow responsible.

Strong winds are also found associated with objects called protostars, huge warm gas balls that have not yet become full-fledged stars in which their energy is provided by nuclear reactions. Recent radio and infrared observations of H_2 and CO molecules in the Orion Nebula have revealed clouds of gas expanding outward at velocities approaching 100 kilometers per second. Furthermore, high-resolution, very-long-baseline interferometer observations have disclosed expanding knots of H_2O maser emission near the star-forming regions in Orion, thus linking the strong winds to the protostars themselves. The specific causes of these winds remain unknown, but if they generally accompany star formation, astronomers will have to consider the implications for the early Solar System; after all, the Sun was presumably once a protostar too.

All things considered, a most fascinating feature of stars is their remarkable repertoire of activity, which displays variations throughout their long lifetimes, from the youthful protostellar stage to the mature stages represented by luminous cool stars. Such activity has been a revelation in recent years; magnetic fields, photospheric violence, flare-like effects, coronal holes, and stellar winds are part of the normal daily routine of essentially every type of star. The atmospheres of most stars are far removed from the quiescent equilibrium state implied by their steady visible image burning forth in the evening sky, and we learned this only by studying their invisible radiation.

Invisible radiation is the key to understanding stellar activity because active regions of stars have densities too low and temperatures too high to give off much visible light. Radiation emanating from their superheated gas is predominantly of the short-wavelength ultraviolet and x-ray type, while the escape of their eventually cooled gas is often accompanied by the emission of long-wavelength radio and infrared waves. Galileo would have been amazed.

Role of Magnetic Fields

Magnetic fields lie at the heart of solar and stellar activity. Such fields seem ubiquitous throughout the Universe; planets, stars, and galaxies are all magnetized to varying degrees. In the case of the star we know best, the Sun, it seems that the magnetic field is produced deep in its interior. It then migrates to the surface, here and there emerging to form the sunspots that puzzled Galileo centuries ago. Only now, with instruments capable of detecting invisible radiation from superhot gas, do we realize that coronal eruptions and flares can trace their energy to that stored in magnetic fields which have emerged from below.

The origin of magnetism inside the Sun and stars poses a fundamental problem to physics. Some astronomers contend that magnetic fields are trapped inside stars at the time they form from interstellar gas. According to this idea, the field then slowly migrates to the surface of the star, occasionally producing sunspots. However, this theory has difficulty explaining why the Sun's magnetic field varies in an 22-year cycle. Moreover, evidence is accumulating that other stars undergo similar cycles, making them difficult to explain as well.

We can look to the Earth for insight, for it too has a magnetic field, as anyone who has used a compass knows. It is easy to show that any magnetism trapped in the Earth when it formed must have decayed long ago, so that the terrestrial magnetic field that we now measure must be continually regenerated in the interior of our planet. Theorists propose that the flow of electrically conducting fluids in Earth's core can do the trick, because such flows generate a potential difference when they move perpendicular to a magnetic field, much like that in an automobile generator. This potential difference can drive a current that regenerates the original magnetic field, but we can prove mathematically that only certain types of flow will work.

Specifically, what is required is both a rotational motion (as is surely present in the Earth) and also a rising and falling motion of the type expected when the fluid is heated from below. Such a flow is called convection; it occurs everytime we heat water on the stove. Likewise, it occurs in the core of the Earth, because heat is generated by the radioactivity of the rocks there. Since convection and rotation of a core made of liquid iron is almost surely responsible for the maintenance of Earth's field, a combination of rotation and convection in the Sun should suffice to maintain its magnetic field. Indeed, the gases of the Sun are observed to rotate (recall Galileo and the sunspots), and also to convect because of the incessant heating by nuclear reactions in the Sun's interior.

To fathom the generation of magnetic fields in stars requires that we have a solid understanding of the circulation of the gases inside our Sun. Since younger stars display higher levels of activity, we infer that they have stronger magnetic fields, and furthermore, we may suppose that these greater fields are in turn caused by the fact that young stars are observed to rotate much faster than the Sun. However, it remains a mystery why some stars more massive than our Sun have among the strongest magnetic fields. Not only do such stars rotate only sluggishly, but we also have reason to suspect that they have no convective flow at all inside. Thus the origin of stellar magnetic fields is by no means a closed subject.

As for the violent activity in the Sun and stars, the often explosive dissipation of magnetic fields and the accompanying acceleration of particles to high energies has become a fascinating problem in astrophysics. Laboratory and theoretical research, as well as magnetic phenomena observed on the Sun and planets, strongly suggest that the "reconnection" of broken magnetic-field lines is a key mechanism. In this mechanism, the topology of magnetic lines of force is rearranged so that two lines which were originally distinct now reconnect, one end of one line joining the other end of the other, and *vice versa.* This process seems to be responsible for the evolution of a relatively quiescent, large-scale motion which transforms magnetic fields into an impulsive, energetic, small-scale phenomenon like a solar flare in which energy is extracted from the magnetic field and used to accelerate particles to very high velocities. We suspect that instabilities in the magnetic field configuration play an important role in triggering magnetic reconnection, and thus, the conversion of magnetic energy into the energy of fast particles.

A better appreciation for magnetic reconnection and particle acceleration is essential not only for solar and interplanetary physics, but also for many other types of violent events in the Universe. This seems especially true of the x-ray sources, the radio galaxies, and the quasars to be discussed in

Chapter 6. Accordingly, studies of the Sun are of importance, not only because our parent star is of interest to us in its own right, but also because only in this relatively near by object can we observe the relevant violence on the scale and in the detail needed to genuinely understand the mechanisms at work in vastly more energetic events far into the Galaxy and beyond.

Prospects for the Future

How can we achieve a reasonable understanding of the exceedingly complex activity of the Sun and stars? On the one hand, the basic equations of physics – the familiar ones of Newton, Maxwell and others – give a complete and accurate description of the behavior of the matter and magnetic fields known present in stars; so far we have not been forced to cast about for new laws of physics. On the other hand, these equations are not solutions; they are only the starting point from which the mathematical solutions must be drawn, and it is up to us to construct those solutions. Unfortunately, given the huge numbers of atoms in a typical atmosphere as well as fields that tend to vary sporadically, the variety of possible solutions is staggering. Thus far, we have made little headway in anticipating *a priori* the various classes of solutions that are likely to be important in real stars.

We must therefore rely to a large extent on an empirical approach, first observing each phenomenon of interest in sufficient detail to recognize the basic physical processes underlying the phenomenon. Then we can construct solutions capable of explaining such phenomena in terms of the fundamental laws of physics. This is the part of the analysis in which solar observations play so crucial a role, for in many cases only the Sun allows observations with sufficient resolution and sensitivity to permit us to identify and study the physical events involved.

Much of our current knowledge of the Sun's activity is limited by lack of good spatial resolution. Our best ground-based observations are usually limited to 1 arc second resolution, which means that we can distinguish solar surface features no smaller than about 700 kilometers across; yet theorists speculate that the most interesting phenomena are only 100 kilometers across. If we are to decipher the fine details, especially in the rather localized sites of intense activity, then better resolution is clearly needed.

To correct this deficiency, two satellites are now being designed. One, a precise 1.3-meter instrument called the *Solar Optical Telescope,* is due to be launched by NASA's *Space Shuttle* during the 1980s; it will be able to resolve features as small as 0.1 arc second, or 70 kilometers on the solar surface. Using this instrument to make both optical and ultraviolet observations, we should see for the first time the fine structure of the magnetic fields as well as the associated motions of gases and heating of small regions.

Somewhat later in the decade, a highly sophisticated *Advanced Solar Observatory*, outlined in Appendix E, will be placed into Earth orbit, where it will be able to image simultaneously the optical, ultraviolet, and x-ray emissions of the Sun with a resolution better than 0.1 arc second. These kinds of close-up simultaneous observations should help to reveal the basic physics of a great variety of phenomena, including flares, sunspots and their surrounding regions, and the activity associated with the ever-changing magnetic fields. Because of the tremendous temperature differences as we pass from the photosphere out to the corona, simultaneous observations in all three spectral ranges are needed to establish the connections between phenomena as they affect the material at different altitudes in the Sun's atmosphere. Such observations will provide tests of present and future theoretical ideas that are usually much more stringent than can be furnished by the best observations from the ground.

The desirability of obtaining extremely detailed observations of the solar surface raises an exciting possibility – an unmanned spacecraft orbiting, or even sent into, the Sun itself. NASA has proposed just such a bold mission, called *Star Probe*, designed to undertake the first *in situ* exploration of any star. Such a craft should be able to safely move to within two million kilometers of the Sun's surface; this may be compared to Earth's 150-million-kilometer distance from the Sun. Even if the craft were to carry only a small 10-centimeter telescope, we would be able to resolve details of solar activity down to 10 kilometers. Unquestionably, such a mission would open up a whole new era of investigation in solar physics, much as NASA's program of planetary exploration by deep-space probes has broadened and supplemented the traditional field of planetary astronomy.

Studies aimed at elucidating the interior structure of the Sun will also receive renewed impetus during the 1980s. About a decade ago, the subject of "neutrino astronomy" was born to probe the deep interior of the Sun. Neutrinos are invisible, chargeless, and virtually massless subatomic particles that are created as by-products of the nuclear reactions churning away in the Sun's core, which of course is hidden from our gaze in all electromagnetic bands. Extraordinarily elusive, neutrinos interact very little with material substances; it is estimated that they can penetrate several light-years of lead without stopping. Although hard to believe, experiments have also shown that one can nonetheless catch at least a few neutrinos in a tank containing tetrachloroethylene (C_2Cl_4), a common cleaning fluid used in the garment industry. Accordingly, researchers have for some years operated a swimming pool-sized "neutrino telescope" using some 100,000 gallons of C_2Cl_4 as a detector; to guard against spurious events generated by other cosmic sources or by unrelated experiments in nuclear laboratories, the entire device is operated deep below the surface in a goldmine in Homestake, South Dakota, where the overlying rock should stop unwanted particles. Solar neutrinos have actually been detected by this novel telescope, but the rate of detection is only about a third of that predicted by current theories of stellar energy generation. This result has forced theorists to reexamine their models of the Sun's interior but the discrepancy remains. If we build a new detector using the rare chemical element gallium instead, it will be sensitive to the full range of neutrino energies generated in the Sun. Such an experiment should not be sensitive to the details of the solar interior as is the present experiment, but simply to whether the Sun's energy originates in nuclear reactions as has long been postulated. Failure to detect neutrinos with gallium would cause a major upheaval in astrophysics, as explained in more detail in Chapter 8, and a new era of science would begin.

In a similar vein, the recent discovery of solar oscillations offers another opportunity to further our knowledge of the Sun's interior. In the 1970s, researchers found that the Sun vibrates with a period of about 5 minutes, and furthermore that these oscillations, like earthquake vibrations on our planet, can be used to probe the Sun's intermediate layers between its atmosphere and core. We anticipate that long-term observations of these vibrations will tell us much about the interior of the Sun, just as earthquakes enable scientists to learn about the interior of the Earth. Recent observations of these oscillations from a station in the Canary Islands have hinted at the remarkable result that the core of the Sun is rotating roughly twice as fast as the surface layers. If confirmed, this finding would have major implications for understanding the Sun, particularly the origin of its magnetic field. It is apparent that such observations must be continued and extended in the future.

Even while we mount a coordinated attack on problems of solar activity, observations of activity in other stars will also be pressed forward with ground-based equipment as well as x-ray and ultraviolet telescopes in Earth orbit. In particular, the *Space Telescope* with its accompanying far-ultraviolet spectrograph mission (*cf.*, Appendix A) and the *Advanced X-ray Astrophysics Facility* (Appendix D) will have far greater sensitivity and resolution than their predecessors and will therefore be able to study far more and varied stars than have been possible with earlier instruments. Our goal is to observe an array of stars having many different characteristics, in order to test the dependence of activity on, for example, the age and spin rate of a star predicted by different theoretical models.

The new instruments of the future will also have powerful capabilities to investigate strong stellar winds and their consequences. The *Advanced*

X-ray Astrophysics Facility will undertake sensitive surveys and spectral observations of x-ray emission from the winds of luminous hot stars, permitting studies of the properties of the coronal gas and the mechanisms for heating it. Ultraviolet sensors aboard the *Space Telescope* will greatly increase the number of observable stars and thus furnish better clues to the physical conditions in the winds of the hottest stars, their driving mechanisms, and their interaction with their interstellar environments.

Infrared observations with new equipment should tell us much about the winds of both hot and cool luminous stars as well as the winds associated with regions of star formation. Because of its high sensitivity, the *Shuttle Infrared Telescope Facility* (*cf.*, Appendix C), among other newly proposed instruments, will permit observations of stellar winds and their associated dynamics in the optically invisible regions where stars normally form. Moreover, the *New Technology Telescope*, a proposed huge new ground-based optical/infrared reflector described in Appendix F, will make possible high-precision measurements of the spectral signatures of atoms and molecules, thereby permitting us to inquire how mass loss affects the evolution of stars and protostars.

Unquestionably, the next decade or two of solar and stellar physics spells excitement, as we push forward our studies of invisible radiation much as astronomers before us have studied the visible Sun and stars. No doubt the Sun will continue to be our most accessible experimental window into the vast array of stellar activity on stars too distant to be resolved.

In another direction, solar activity, while exceedingly complex, offers a laboratory to study plasma processes occurring in a variety of other astronomical objects from the interstellar medium to quasars. With its tremendous temperature range from its photosphere at 6000 degrees, to its corona at 1,000,000 degrees, to the hearts of solar flares at 100,000,000 degrees, our Sun provides us with the opportunity not only to study solar phenomena that affect our earthly abode, but also basic plasma processes occurring throughout the Universe.

Even so, and despite its recently discovered repertoire of violence, we can be thankful that our Sun lies at the low end of the range of cosmic activity. The fact that we, as life forms, are associated with an only slightly active star might not be coincidental. It seems doubtful that life could arise in the vicinity of objects that are much more active.

Speculation Run Amuck

As a matter of personal educational policy, I usually spend a few minutes at the start of each class discussing a recent discovery – some new item of information either announced in the current journals or, more often than not, revealed during a private conversation. Once, when I was a Harvard faculty member during the late 1970s, my desire to keep my students informed of such late-breaking events led to some heavy speculation that spread like wildfire through Cambridge and beyond.

Caltech astronomer Charles Kowal had optically detected a new celestial object where none had apparently been previously seen. As I told my class of five hundred Harvard and Radcliffe students, various technical reasons implied that the object resided in our Solar System, but it was unlikely to be a comet since it displayed no tail or other fuzziness normally associated with such a rapidly evaporating ball of dirty snow and ice. Nor could the new object be a conventional asteroid, for its distance seemed to range between Saturn and Uranus, well beyond the well-known asteroid belt between Mars and Jupiter. My students were fascinated, as was I, that even in the late twentieth century we are still taking inventory of new heavenly bodies so close to home. Needless to say, with the movie *Star Wars* having been released just a few weeks prior, some guy in the audience yelled, "It's Darth Vader on the way!" after which everyone had a good chuckle and I resumed my scheduled lecture.

As is often the case, the Caltech group had telephoned news of the discovery to astronomers at the Harvard Observatory in order to check the relevant data stored in the Harvard Plate Stacks, a virtual library of photographic plates that extend back to the birth of astronomical photography some hundred years ago. None of these older photographic records showed evidence for the new object, but one of the Harvard astronomers, William Liller, decided to confirm its existence by taking a fresh series of photographs with a nearby telescope at the old Agassiz Field Station in the village of Harvard, Massachusetts. I remember eagerly awaiting his return, for I wanted to be able to share with my students the latest findings of this intriguing piece of astrodetective work. But when the plates were developed, something startling occured: *two* new objects were now present in the field of view! Whereas Kowal had made an initially rare find, Liller had beaten the astronomical odds by sighting yet a second new object in the same region of the sky.

The next day I announced these new data to my class, and, just as I expected, the same Harvard man quipped, "Darth really *is* on the way!" Of course, no one believed him, though no one chuckled this time either. Indeed, we had witnessed an extraordinarily rare event, for the odds must be greater than astronomical that a given region of the sky, virtually immutable for at least a century, would rather suddenly spawn two new cosmic objects on successive nights.

Observatories everywhere were alerted by Smithsonian's Central Bureau for Astronomical Telegrams, and astronomers around the world used telescopes on the ground and in orbit to monitor these peculiar happenings. Observations during the next few nights resolved the issue. The first new object seems to be a 100-kilometer-sized maverick asteroid (or escaped moon of Saturn) that revolves about the Sun every 50 years in an orbit ranging from 8 to 20 astronomical units; named Chiron after one of the centaurs of Greek mythology, it will make its next closest approach to Earth in 1996. The second new object was proved to be a supernova that burst forth in a faraway galaxy that coincidentally just happened to be at that time in the same field of view as Chiron. The distant galaxy could be discerned only with a long time exposure, after which it became clear that the second new object had no relation at all to the first; to be sure, the supernova was millions of light-years beyond our Solar System.

This is a good example of how careful observations can objectively decipher what at first seems bizarre and often causes much subjective speculation. Indeed, the ultimate reliance on the authority of the experimental test is the one feature that clearly distinguishes the scientific enterprise from all other ways of explaining nature. As the British astrophysicist, Sir Arthur Eddington, once declared, "For the truth of the conclusions of physical science, observation is the supreme court of appeals".

E.J.C.

4 Planets, Life, and Intelligence

Are We Alone?

On the 7th day of January in the present year, 1610, in the first hour of the following night, when I was viewing the constellations of the heavens through a telescope, the planet Jupiter presented itself to my view, and as I had prepared for myself a very excellent instrument, I noticed a circumstance which I had never been able to notice before, owing to want of power in my other telescope, namely, that three little stars, small but very bright, were near the planet; and although I believed them to belong to the number of fixed stars, yet they made me somewhat wonder, because they seemed to be arranged exactly in a straight line, parallel to the ecliptic, and to be brighter than the rest of the stars, equal to them in magnitude. The position of them with reference to one another and to Jupiter was as follows.

On the east side there were two stars, and a single one towards the west. The star which was furthest towards the east, and the western star, appeared rather larger than the third.

I scarcely troubled at all about the distance between them and Jupiter, for, as I have already said, at first I believed them to be fixed stars; but when on January 8th, led by some fatality, I turned again to look at the same part of the heavens, I found a very different state of things, for there were three little stars all west of Jupiter, and nearer together than on the previous night, and they were separated from one another by equal intervals...

At this point, although I had not turned my thoughts at all upon the approximation of the stars to one another, yet by surprise began to be excited, how Jupiter could one day be found to the east of all the aforesaid fixed stars when the day before it had been west of two of them; and forthwith I became afraid lest the planet might have moved differently from the calculation of astronomers, and so had passed those stars by its own proper motion. I therefore waited for the next night with the most intense longing, but I was disappointed of my hope, for the sky was covered with clouds in every direction...

Accordingly, on January 11th I saw an arrangement of the following kind, namely, only two stars to the east of Jupiter, the nearer of which was distant from Jupiter three times as far as from the star further to the east; and the star furthest to the east was nearly twice as large as the other one; whereas on the previous night they had appeared nearly of equal magnitude. I therefore concluded, and decided unhesitatingly, that there are three stars in the heavens moving about Jupiter, as Venus and Mercury round the Sun; which at length was established as clear as daylight by numerous other subsequent observations.

From *The Sidereal Messenger*
by Galileo Galilei, Venice, 1610

Molecular Biology continues to give powerful insight into the nature of life. Complex DNA (deoxyribose nucleic acid) molecules in the heart of every cell encode all the information required to determine how that cell develops and functions. All the rich diversity of life on Earth is coded in strands of DNA, which have evolved from primitive forms that apparently arose through a series of chemical reactions in our corner of the Universe. Studying the myriads of stars in the Universe, we wonder what other genetic systems might exist.

Are we alone in the Universe? Do planets exist around other stars? Might there be astronomers on alien worlds petitioning their governments to support their exploration of the cosmos? Could there really be a galactic civilization, composed of a network of planetary life forms, linked by powerful communications beams? The prospects for extraterrestrial life have been pondered since ancient times, but only in our generation have earthlings gained the technological tools needed to investigate the subject meaningfully.

The Solar System

The National Aeronautics and Space Administration's program of Solar-System exploration by robot spacecraft is one of the grand technological and scientific adventures of our generation. The opportunities to reconnoiter and sometimes to orbit planetary bodies, to study their properties at close range, to make detailed measurements of their atmospheres, magnetic fields, and moons, and in some cases actually to land upon their surfaces, have given birth to the new field of planetary science. This is truly an interdisciplinary subject that draws on a number of scientific endeavors, and relates strongly to the rest of astronomy and astrophysics.

For example, studies of planetary atmospheres have led to the development of concepts and techniques that apply more broadly to the dynamics of the atmospheres of stars and to interstellar matter; *in situ* measurements of planetary magnetism have given birth to new theories explaining how particles can be accelerated in magnetic fields to nearly the speed of light; evidence for the internal heating of Jupiter-like planets bears strongly on the general issue of the contraction and structure of protoplanets – the bodies that must have preceded the planets as we know them; observations of comets, meteorites, and other primitive objects of our Solar System have provided insights into the chemical composition of the primordial nebula from which the Sun and planets originated, and these studies in turn should apply to the formation of stars and planets from interstellar clouds throughout the Galaxy.

Equally significant, by studying the other planets, we are gaining a deeper understanding of the many natural phenomena occurring within the Earth, in its atmosphere, and in its oceans. In the newly emerging concept of "comparative planetology", the Solar System is considered a giant laboratory where many of the natural phenomena indigenous to Earth are also recognized to occur in other planetary settings, but under diverse physical conditions and often at different evolutionary stages. Although we are only now in the early stages of development of this branch of science, we anticipate that broadening our experience to such a divergent range of planetary conditions will lead to a keener understanding of the rock that is our home in space.

The most clearly recognizable goals in planetary science today are two: to attain a balanced and more complete exploration and theoretical interpretation of the current state of the planets and their satellites, and to establish a broad empirical basis for deciphering the origin of our Solar System.

Aggressive efforts to address the first of these goals led to some spectacular advances in the 1970s, as novel research techniques virtually revolutionized our knowledge of the Solar System. Spacecraft reconnoitered three planets (Mercury, Jupiter, and Saturn) and a host of peculiar moons for the first time, successfully landed and operated on the surface of both Mars and Venus, directly probed several locations in the thick atmosphere of Venus, and returned samples of material from several sites on the Moon. In addition, an array of increasingly sophisticated ground-based instruments gathered data that complement and extend the findings of each of these deep-space missions. Consider just a few of the scientific highlights stemming from these efforts.

At Mercury, a *Mariner* spacecraft recorded surface features – such as craters – that point to the same ancient heavy bombardment that scarred our Moon and much of the surface of Mars. Signs of evolution in Mercury's interior show clearly in its own special brand of surface features: widespread volcanic plains – quite different from those on the Moon – as well as huge planetary streaks, implying that internal melting, cooling, and shrinkage once plagued this alien world. Mercury's magnetism was an unexpected discovery, given its sluggish rotation and the current theories for the origin of magnetic fields discussed in Chapter 3.

Radar signals bounced off the planet Venus by American scientists and television signals broadcast by the Soviet *Venera* spacecraft gave us our first glimpse of its surface terrain. The radar observations, penetrating the clouds enshrouding the planet, reveal a rugged surface complete with mountain ranges, canyons, plateaus, and possible volcanoes, although there are no indications yet of Earth-like crustal plates similar to terrestrial continents. Confirming these properties, several spacecraft were able to measure the natural radioactivity of surface rocks before succumbing to the extraordinary heat of Venus' surface (~450 degrees Celsius); the results are so close to those of several terrestrial rock types that we can assume that on Venus, as on Earth, mountain building has taken place. As for the Venusian atmosphere, observations made both from Earth and from several *Pioneer* entry vehicles prove that the upper layer of clouds is made mostly of concentrated droplets of sulfuric acid (H_2SO_4). The atmosphere, which is composed largely of CO_2 with smaller amounts of nitrogen and water, may have been released from volcanoes during more active periods in the past.

At Mars, the *Viking* and *Mariner* missions added much to our knowledge of the evolution of planets and their atmospheres. For example, we now have clear evidence that volcanic and other mountain-building events occurred early in Martian history; yet, wide expanses of the planet also show primitive cratered landscapes untouched except for atmospheric erosion. Some of these craters have forms suggesting that a large reservoir of water lies beneath much of the surface in the form of permafrost – frozen ground of the same sort that underlies many northerly areas on Earth. Also found on the surface are relics of what are almost certainly stream and flood channels (although no liquid water is present now), implying significant flooding and a more clement climate during Mars' early history. The red planet probably experiences long-term climatic variations (like extended ice ages and interglacial warming trends) as the planet orbits the Sun. Indeed, measurements of the relative abundances of some of its atmospheric gases suggest that Mars probably once had a denser atmosphere, though a exact chronology of Martian history still eludes us.

Our newly acquired data about the makeup of the Martian and Venusian atmospheres, in conjuction with our more detailed knowledge of Earth's atmosphere, marks the beginning of our efforts to construct comprehensive models of atmospheric evolution for the inner parts of the Solar System. For example, direct measurements of meteorological conditions on Mars' surface have led to an increased understanding of the seasonal Martian variations that have long been observed from Earth; the details of Martian dust storms and the overall circulation of its lower atmosphere turn out to have considerable similarities to such processes on Earth. Also, the curious Y-shaped markings seen in ultraviolet images of Venus' upper cloud deck seem to be driven by a global system of atmospheric waves possibly akin to Earth's jet stream, although continuous buffeting by the solar wind makes it hard to unravel the specific mechanisms responsible; whatever the process, it may eventually help us better understand the upper-level weather patterns that seemingly dominate the spread of high and low pressure zones on our own planet. And of some future significance, perhaps, is the fact that the stratosphere of Mars displays an ozone chemistry closely analogous to what would be expected on a highly polluted Earth.

In the realm of the outer planets, the spectacularly successful flybys of the *Pioneer* and *Voyager* spacecraft through the Jupiter and Saturn systems provided a great leap forward in our study of these truly huge worlds. In particular, we now have a much sounder knowledge of their chemical abundances, having found a wide variety of hydrogen and carbon compounds on Jupiter, Saturn, and Saturn's moon Titan. Also discovered were substantial deviations from simple equilibrium conditions in the deep atmospheres of those two planets; vertical mixing of hot invisible regions with the cold visible layers must surely be an important process. Infrared spectra obtained from Earth, as well as close-up observations from spacecraft, revealed exquisite details about the deep cloud structures on Jupiter, its giant atmospheric eddies, and its surprising auroral activity, including the sighting of huge lightning flashes (noted in the visible band by *Voyager* as well as in the ultraviolet band by the *International Ultraviolet Explorer* satellite in Earth orbit).

Images of the incredibly complex cloud belts, whirls, and spots on Jupiter and Saturn, telemetered back to Earth during the close encounters with our spacecraft during the late 1970s and early 1980s, aroused wonder among lay persons and scientists alike.

As for the outermost giant planets, we still know little about their principal properties. Even our best ground-based methods of determining planetary spin give embarrassingly divergent results for the rotation periods of Uranus and Neptune. Should the *Voyager* spacecraft, now scheduled for a 1986 rendezvous with Uranus, survive the rigors of deep space, it will no doubt radio to Earth some spectacular data regarding this still mysterious planet; the Uranian gravitational field is also slated to "sling-shot" *Voyager* in the direction of Neptune, where it will glide past this equally enigmatic planet in 1990. We can only hope that *Voyager's* electronic circuitry is up to the task nearly two decades after its launch from Earth.

Studies of rings and satellites of planets came into their own during the fast-paced 1970s. Both spacecraft and ground-based instruments were involved in the discovery of ring systems around Jupiter and Uranus, leading some researchers to suggest that rings might be common by-products of the events that create systems of satellites around the planets. Several new satellites were also recently sighted, including one orbiting the remote planet Pluto, as was the first member, Chiron, of a new class of small asteroids* beyond Saturn. These discoveries aside, surely Galileo himself would have been astonished at the surface streaks, the strange atmospheres, and even the active volcanos discovered on the Jovian satellites that he first saw one night in the year 1610. Indeed, *we* were astonished when *Voyager*, gliding past these alien worlds in the year 1981, radioed back these unexpected data.

Origin of Planetary Systems

The second major goal of modern planetary research concerns the origins of planets and their satellites. On the basis of what we now know about interstellar matter, we can legitimately argue that the mechanisms that led to the formation of the Sun and its planets must be at work throughout the Galaxy today. Much that we are now learning about the formation of our Solar System, both from the NASA planetary-exploration program and from ground-based planetary astronomy, can thus be applied to the more

* An asteroid is a small planetoid orbiting the Sun; Chiron's diameter is a hundred kilometers.

general problem of planetary formation. Even so, we know of no such planets outside our Solar System. Over the past few decades, observers have occasionally reported that some nearby Sun-like stars seem to move ever so slightly to and fro, implying that those stars are affected by the gravitational pulls of unseen accompanying planets. But, because these reports have not yet been unambiguously confirmed, we must admit that we have no conclusive evidence that any planets exist beyond our Solar System.

Current theories of star formation suggest that a fragment of a dark interstellar cloud destined to become a star would have some spin, thus setting into rapid rotation the gas and dust of the cloud-to-be-star while inwardly contracting under gravity's influence. The resulting centrifugal force would eventually resist the inward pull of gravity and cause the spinning disk of gas and dust to become a natural breeding place for planets, arrayed like ours, in a great flattened and rotating system.

The connection between interstellar clouds and our primitive planetary system was brought out forcefully when, in 1973, astronomers used radio techniques to detect several rather complex molecules in Comet Kohoutek, including methylidyne (CH), hydrogen cyanide (HCN), and ionized water (H_2O^+). Suddenly, comets, once regarded by some astronomers as useless cosmic debris, became another "Rosetta Stone", providing a direct interface between science of the Solar System and that of interstellar matter. Thus we now have a relatively local "laboratory" in which to study the molecules that perhaps could have survived the contraction phase of the original presolar cloud to form the Solar System. In the coming years, we need to clarify to what extent small bodies (*e.g.*, comets, asteroids, meteorites and interplanetary dust) represent unmetamorphosed primitive material, as has long been speculated.

Specifically, in the case of our Solar System's formation some 5 billion years ago, we conjecture that dust particles (either interstellar particles brought in when our parent nebula first formed or particles that condensed later from gases in a heated nebula) migrated to its midplane, whereupon their mutual gravitational attraction drew them together to form moon-sized solid bodies which in turn accumulated to form the inner Earth-like planets and the cores of the outer Jupiter-like planets. This model has been severely constrained by the discovery of an excess of the isotope magnesium^{26} in almost pure alumina crystals in meteorites, which are bodies thought to be fragments of asteroids. This isotope is the daughter of radioactive aluminum^{26}, which has a lifetime of only a few million years. In order for the magnesium^{26} to be segregated in alumina crystals, which are made of the stable isotope of aluminum, namely aluminum^{27}, the alumina crystals had to have formed before the aluminum^{26} decayed. Hence the radioactive aluminum found its way from the supernova explosion (where it must have formed) into a solid body within a few million years. All this implies not only that the formation of the Solar System must have occurred faster than previously thought possible, but also that large amounts of aluminum^{26} must have been injected just before the process started. Since both supernova explosions responsible for producing heavy elements such as aluminum^{26} and the origin of planetary systems are relatively rare events, we suspect that they were causally related. Indeed, the notion that our Solar System might have formed because the concussion of a supernova triggered the contraction of our parent molecular cloud is now gaining support as observers continue to find appropriate alignments of young stars in the vicinity of other supernova remnants in our Galaxy.

But these ideas are speculative. To prove that planets exist around other stars, we must observe them. The most straightforward method, to search near a star for a speck of starlight reflected from a planetary companion, is not currently feasible. Even a planet as large as Jupiter reflects only a billionth of its parent star's light, and such a speck of light would be lost in the glare of the parent star. Another method is indirect, based on the fact that Jupiter's gravitational pull would vary the Sun's radial velocity by a few tens of meters per second, and, to a distant observer some 10 light-years away, would cause a thousandth of an arc second displacement (or wobble) of the Sun's position. Improved observational and data-reduction techniques have only recently become available to make velocity and positional measurements of the required precision. What we now need are sustained observations of many candidate stars that could possibly reveal alien planetary systems, provided they really exist. Presumably, if they are found, great effort would be expended to detect their reflected light with the powerful telescopes of the future. Because of its sharp images, *Space Telescope* would be the instrument of choice, but the task would be formidable even for that great device.

Life in the Solar System

Life on Earth is so intertwined with the chemistry of our atmosphere and oceans as to form a single ecosystem in which each part is affected by the others. Based upon one of the most abundant elements in the Universe, namely carbon, and dependent on a readily available liquid, namely water, the ecosystem is sustained by the continual capture of our Sun's energy by photosynthesis in plants. Using these simple ingredients, primitive life forms managed to evolve over billions of years by mutation and natural selection into new species of ever-increasing complexity. Beyond these basic considerations, we are uncertain what other properties of planet Earth played essential roles in the origin and development of life and intelligence. A solid surface? The pull of gravity? The daily cycle of light and dark? The monthly cycle of lunar tides?

Astronomical research has demonstrated the existence of vast numbers of stars similar to our Sun. Our efforts have also proved that the abundances of chemical elements are much the same everywhere. However, while we concede the possibility that habitats for life might be scattered throughout the Universe, the assumption that biochemical events operate wherever the conditions are favorable remains speculative until other examples of life in the Universe are actually found.

Several planets in our Solar System are similar enough to Earth that a search for life on them can shed some light on these issues. For example, Mars has long appealed to scientists and the public alike as an intriguing target for exploration. The 1970s have seen a remarkable effort in that direction, culminating in the orbiting spacecraft and landers of the *Viking* missions. Although the most well-known results were the photographs of the terrain around the landing sites and the chemical searches for microorganisms, the landed and orbiting spacecraft radioed back to Earth a huge amount of data regarding the meteorological conditions of the Martian atmosphere as well as the geological processes that have shaped the Martian surface. The conclusion that has emerged from these studies is that the Martian atmosphere, now dry and cold, was perhaps quite different in the past; probably large amounts of water precipitated at one time, scarring the surface with torrential floods. Although the *Viking* landers were unable to find any organic matter at the two surface sites sampled, it is not out of the question that conditions about a half-billion years ago were more favorable to life. Ironically, while Mars may now be biologically dead, it may someday become an archeological bonanza. Only further exploration can tell for sure.

Venus has been extensively studied by both the United States and the Soviet Union. Two straightforward facts discovered by our spacecraft have made it clear that Venus is inhospitable to life: Its surface is hot enough to melt lead, and its atmosphere is topped by clouds laced with liberal quantities of sulfuric acid. Mercury is also too hot, and airless as well, both properties caused by its proximity to the Sun.

Jupiter continues to elicit speculation that living organisms or at least pre-biotic substances might be present because its dense atmosphere contains organic molecules that likely represent some of the first steps toward the origin of life. Although the temperature is low at the visible cloud tops, it increases inward; thus there must be some atmospheric depth at which organisms could be quite comfortable, though we find it difficult to imagine how they would avoid being cooked by falling or frozen by rising. Furthermore, we have no clear evidence for a solid or liquid surface to which life might cling – if, indeed, that is essential. As for the Jovian moons, only Io seems to have sufficient heat and atmosphere for life as we know it, but the *Voyager* spacecraft also proved that this object is overrun by hot lava and poisoned with sulfurous fumes now gushing from several active volcanoes, making its surface undoubtedly hostile to most known forms of terrestrial life.

In the outer realms of our Solar System, most objects are too cold for life to be credible, although Titan, the largest moon of Saturn, is now known from *Voyager* observations to have an atmosphere composed mainly of nitrogen (much like the Earth's). The temperature at the surface is a chilly − 180 degrees Celsius, close to the temperature at which methane can be either solid or liquid. It is conjectured that there might be a deep ocean of liquid methane on the moon Titan, with polar caps of frozen methane much like polar ice caps on the Earth. About 1% of Titan's atmosphere is methane gas, and small amounts of hydrogen cyanide (HCN), cyanoacetylene (HC_3N), and cyanogen (C_2N_2) have been detected. These molecules are an exciting discovery, for biochemists find that they are among the building blocks of complex organic molecules that are the basis of Earth life. They may be the result of chemical reactions between methane and nitrogen gas (N_2) in the atmosphere of Titan; it is conceivable that such reactions are producing even more interesting organic molecules, which may be detected in the future. While life itself is unlikely on Titan, this richly organic moon may well house a chemical evolutionary laboratory of great importance to studies of life's origin.

Searching for Extraterrestrial Intelligence

In all likelihood, life in our Solar System has so far been confined to Earth and possibly Mars. On Earth, the development of intelligence has led to the explosive growth of technology and thus to radio communications and powerful radars; the earliest radio signals generated a half-century ago have by now propagated 50 light-years into space, past several thousand stars more or less like our Sun. Remarkably, the radio and television signals generated copiously on Earth could be detected by faraway civilizations that would otherwise have no way of knowing about the details of the planet that we call home. They may have already done so. Conversely, at least in principle, we might discover any such beings by using similar means. But should we try to do so?

Should the human race search seriously for signals broadcast by other possible galactic civilizations? Much has been written about this issue, both on a technical and a philosophical level. Reception of intelligent signals from space could have a dramatic effect on human affairs, as did contact between the native peoples of the New World and the technologically more advanced peoples of Europe. The effects would be beneficial, provided the information could be deciphered and should prove generally useful; on the other hand, they could be harmful if humanity is unprepared to use the information wisely.

As far as we know, only if an alien planet harbors intelligent life capable of generating electromagnetic signals detectable at Earth do we currently have any hope of finding life beyond our Solar System. We now have the technology to make significant, long-term searches for such extraterrestrial signals. For example, the radio telescope of the National Astronomy and Ionosphere Center at Arecibo, Puerto Rico, is capable of receiving messages beamed at us from any of the hundreds of billions of stars in our Galaxy, provided the civilization sending the message were transmitting with a facility similar to ours at Arecibo. Several brief searches for such extraterrestrial signals have already been undertaken, so far with negative results. But the rate of improvement in communications technology is so rapid nowadays that each search has been far more sensitive than its predecessors. To be sure, by using the most modern equipment available, we could now undertake a search some million times more effective than any of the previous efforts.

Prospects for the Future

Research momentum in planetary science depends critically on an active program of deep-space missions. Prominent among the planned missions is the appropriately named *Galileo Orbiter and Probe* spacecraft scheduled to be placed into orbit around Jupiter sometime in the late 1980s. The Orbiter will be able to focus with unprecedented resolution on the dynamical properties of specific cloud features, especially the Great Red Spot and its many accompanying whirls, while the Probe will descend through Jupiter's thick atmosphere, relaying back to Earth a wide variety of meteorological and physical data.

More sophisticated infrared telescopes on Earth and in Earth orbit will do much to further our understanding of the planets and moons, especially those worlds beyond Mars. By imaging planetary surfaces and spectrally analyzing their atmospheres, new equipment in orbit will be able to provide astonishing detail of the many peculiar planetary properties of which NASA's spacecraft have, until now, given only glimpses. For example, of great importance is a new infrared device now in the early stages of design. Tentatively called the *Large Deployable Reflector* and described in Appendix G, this facility should greatly improve infrared spectral studies of the planets and their moons, as well as of comets and some asteroids. Because of its large diameter (about 10 meters), spatial resolution at a wavelength of 20 microns should be better than 1 arc second. That's about a ten-fold improvement over any infrared equipment now operating on the ground or in orbit, and sufficient to discern infrared details of all but the most distant planets.

The *Space Telescope* described in Appendix A will have a unique capability to image Jupiter and its moons from Earth orbit with 150-kilometer spatial resolution; such resolution has been previously achieved only briefly as *Voyager* glided past during its 1981 close encounter. Additionally, *Space Telescope* will be able to spectrally analyze planetary atmospheres using the Sun's reflected ultraviolet radiation.

This same technique, called reflectance spectroscopy, should provide valuable insight regarding some of the minor bodies of our Solar System. For instance, we shall gain a better knowledge of the complex chemistry, minerals, and ices of comets by using the *Space Telescope* to study the way they reflect the Sun's ultraviolet radiation. Likewise, observations of infrared radiation reflected from the asteroids, which only more advanced equipment can provide, would considerably extend our knowledge of these stray rocks that sometimes come dangerously close to Earth.

A principal objective for many of these newly proposed devices concerns higher spatial resolution, a direct beneficiary of which will doubtless be projects designed to detect planets around other stars. Vigorously pursued, such projects would be relevant not only to the search for extraterrestrial life but also to a range of important issues in galactic astronomy and astrophysics. By upgrading some ground-based facilities to their current technological limits, we should be able to infer the presence of any Jupiter-sized objects orbiting some of the nearby Sun-like stars. Furthermore, without our atmosphere fuzzing the stellar images, additional improvement in positional accuracy, perhaps by as much as several orders of magnitude, would become available with an appropriately designed telescope in orbit, making it possible to infer the existence of planets as small as Earth orbiting stars as far away as 10 light-years.

The rash of discoveries of molecules in space during the 1970s means that the subjects of interstellar and cometary chemistry are here to stay. Of potentially great significance is the fact that many of these molecules are organic in nature, suggesting that chemical processes relevant to the origin of life are ubiquitous. For example, the interstellar molecules include formaldehyde (H_2CO) and hydrogen cyanide (HCN), while the cometary molecules include the same cyanide and several other rich organic species, all of which play important roles in laboratory experiments aimed at recreating the chemical evolution preceding the origin of life.

When a molecular cloud contracts to form stars and planets, it seems likely that its complex molecules would be destroyed, for heat intensifies as a cloud contracts, subjecting the fragile molecules to breakage. On the other hand, some molecules might survive the extreme conditions encountered during collapse by attaching themselves to the dust grains known to populate interstellar clouds. The complex organic molecules found in certain carbon-rich meteorites (called "carbonaceous chondrites") might therefore be of interstellar origin. Such a process could be critically important for the origin of life in our Solar System, since the available carbon atoms would otherwise tend to be locked up in either carbon monoxide (CO) or methane (CH_4) molecules, both of which are too light to be gravitationally bound to planets as small as the Earth. By condensing onto grains, carbon might instead be carried along when dust is collected into planet Earth, where it would later be available to participate in life.

The syntheses of organic molecules in space will be studied more intensively in the years ahead. By enabling observations of new molecules and of different states of already discovered species, the *Submillimeter-Wave Radio Telescope* and the *Large Deployable Reflector,* among other instruments, offer opportunities to pursue the chemistry of carbon in interstellar clouds, and particularly in the collapsing cores of dark clouds where stars and planets most likely form. Of perhaps even greater relevance to the origin of life, these new devices will grant us additional opportunities to study the molecular composition of comets, which many researchers regard as pristine samples of the most primitive matter in our Solar System.

With regard to intelligent life forms in space, we can hardly imagine a more exciting discovery, or one that would have greater impact on human perceptions, than the unambiguous detection of extraterrestrial intelligence. Quite simply, successful contact would be one of the most profound discoveries of this or any age. Even the complete failure to find any evidence, after long and careful searches of many kinds, would have important implications for the significance of life in the Universe. Accordingly, in the 1980s, we shall see the beginnings of a specific, dedicated, multi-year effort to discover extraterrestrial intelligence. These efforts should not be regarded as "projects", but rather as long-term activities that will surely evolve and perhaps even change radically in their search strategies and techniques.

To begin the search, a sophisticated microwave receiver comprising ten million channels is now being constructed. Used at radio telescopes already in existence, this new equipment will permit us to survey the entire sky for artificial signals as well as to listen especially carefully to several hundred Sun-like stars within our cosmic neighborhood. The radio domain has been chosen for the search largely because, at least by our present understanding of technology, some types of powerful transmissions (*e.g.* military radars and radio communications) on which we might hope to eavesdrop are likely to lie in this general part of the electromagnetic spectrum, and furthermore, for a given amount of transmitted power, the received signal strength should be greatest in the 0.3–30 centimeter wavelength range.

Indeed, most astronomers believe that if any contact ever does materialize with extraterrestrials, it will probably not be of a physical nature, but rather an "electromagnetic" contact employing invisible signals such as long-wavelength radio waves. Thus, perhaps profoundly, the notion of "invisibility" in the Universe might pertain to more than inanimate matter. Lurking in the darkness may well be networks of intelligent life forms that ordinarily never "see" one another, but transact their galactic business by means of radio or other invisible radiation.

Clearly, we are entering an exciting era, not only because we now have the technical feasibility to detect planets around nearby stars, but also because we can address aspects of the issue of life's origin with astrophysical methods, and search for signals from intelligent life on planets so far away that we are unable to see the distant planets themselves. Admittedly, our interest in the tiny fraction of Solar-System matter that condensed into planets is heightened by the fact that life has developed on at least one of them. Have planetary condensations and the origin of life occurred elsewhere as well? And has that life evolved into communicative intelligence, with whom we human beings might be able to entertain a conversation about life in the Universe?

These questions reach far beyond astronomy, and even beyond science as we currently practice it. Yet, as astronomers, we are in a sense commissioned by society to keep a watchful eye on the Universe, and thus we feel duty bound to ask them, as well as to examine how we might try to answer them.

Contact?

During the early stages of researching my doctoral dissertation at Harvard, I found myself suddenly confronted with a potentially spectacular diversion. One night, deep into the graveyard shift at the Haystack Observatory, I happened to be monitoring radio signals in the direction of the invisible nebula known as W49. This is a region of young stars (and possibly planets) on the other side of our Milky Way Galaxy, some 40,000 light-years from Earth. Having studied the region with several different telescopes and at several different frequencies, I knew W49 and its environment reasonably well. Invariably, the spectral analyzers showed this region to emit microwave features that grew out of the noisy static like bell-shaped curves. However, one early August evening in 1970 was to be different.

Alone in the control room while going about my duties of tweaking the experimental gear, I noticed nothing odd about the performance of the electronic gear – with one exception: The spectral analyzer was no longer recording bell-shaped curves from the nebula, but was displaying a slightly stronger radio signal which at first showed no apparent pattern. Intrigued, I continued to monitor the alien signal, and only after several minutes of averaging did I discern the pattern shown in frame (d) of the accompanying figure—a "stickman" much like that drawn by children in kindergarten. I quickly did some observing tests suggesting that the signal was not likely originating from some place on Earth, and I tracked the region until it set below the horizon, after which the signal ceased.

Though I knew little at the time about the prospects for extraterrestrial intelligence, I did know that pictorial images of this sort could be one way that advanced civilizations might try to signal neophyte technological civilizations such as ours. I also knew that I needed to confirm the signal before telling anyone, lest I be the laughing stock of the Harvard Observatory. Obviously, the next twenty-four hours were among the

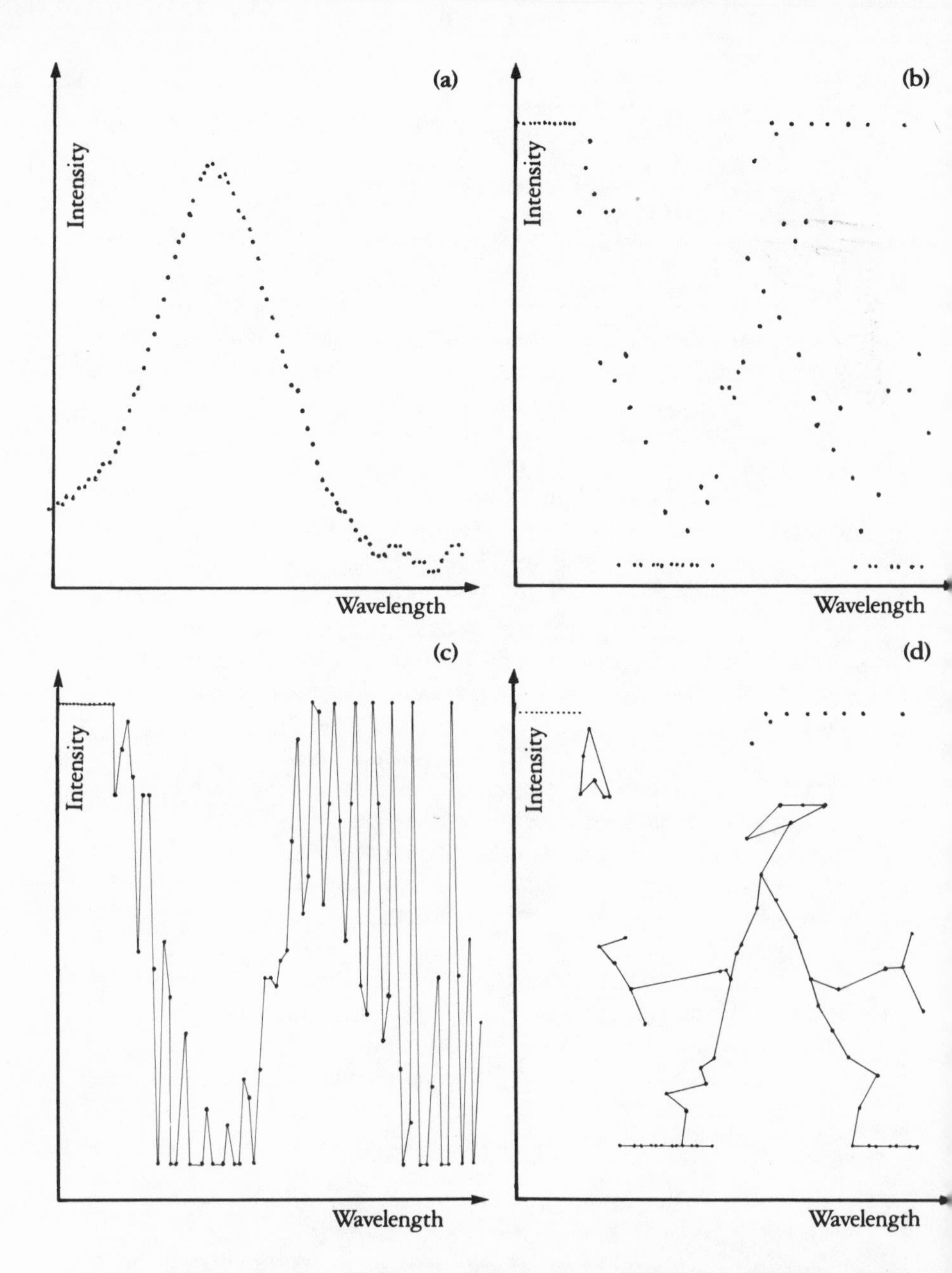

Radio signals normally emitted by the nebula W49 are typified by the bell-shaped curve in (a); this plot represents the intensity (or strength) of an emission feature centered at a frequency of 7792 MHz, and results from electrons changing states within some of the nebula's hydrogen atoms. The peculiar observation described in the text is illustrated in (b). When the adjacent data points are connected, as in (c), the plot takes on the appearance of intense interference. But when the observed data points are connected more imaginatively, as in (d), then a seemingly startling pattern emerges.

most exciting and anxious of my career. Was I going to be the one to make perhaps the most exciting discovery since the birth of language? I was still unsure what I would do next if the signal were confirmed – and if I could completely satisfy myself that it was not a hoax or interference of some sort; after all, Russian radio astronomers have twice announced the discovery of E.T., only to be dismayed by the news that what they really found was America's Big Bird spy-satellite network scanning the Soviet Union.

Making a long story short, I never did manage to confirm the bizarre signal toward W49. But whenever I have access to a radio telescope, I often take a brief glance at the region, hoping to find some corraborating evidence that life exists there or anywhere else in our Galaxy. This may ultimately be the way that life is discovered elsewhere in the Universe – serendipitously as part of some radio astronomer's routine observing program. Indeed, several astronomer-friends have observed peculiar, momentary, and unconfirmed signals that might have originated from a galactic civilization. Only time will tell if any of these apparently spurious events are real, but, in the interim, we must maintain a healthy degree of skepticism while abiding by the experimentalists' dogma regarding new and unexpected findings: guilty until proven innocent.

E.J.C.

5
Galaxies
Nature's Grand Structures

The next object which I have observed is the essence or substance of the Milky Way. By the aid of a telescope any one may behold this in a manner which so distinctly appeals to the senses that all the disputes which have tormented philosophers through so many ages are exploded at once by the irrefragable evidence of our eyes, and we are freed from wordy disputes upon this subject, for the Galaxy is nothing else but a mass of innumerable stars planted together in clusters. Upon whatever part of it you direct the telescope straightaway a vast crowd of stars presents itself to view; many of them are tolerably large and extremely bright, but the number of small ones is quite beyond determination.

And whereas that milky brightness, like the brightness of a white cloud, is not only to be seen in the Milky Way, but several spots of a similar color shine faintly here and there in the heavens, if you turn the telescope upon any one of them you will find a cluster of stars packed close together. Further – and you will be more surprised by this, – the stars which have been called by everyone of the astronomers up to this day nebulous, are groups of small stars set thick together in a wonderful way, and although each one of them on account of its smallness, or its immense distance from us, escapes our sight, from the commingling of their rays there arises that brightness which has hitherto been believed to be the denser part of the heavens, able to reflect the rays of the stars or the Sun.

From *The Sidereal Messenger,*
by Galileo Galilei, Venice, 1610

LIKE THE GALAXY IN WHICH WE LIVE, each of the ten billion or more galaxies in the known Universe must be fascinating systems in their own right. Spinning ever so slowly, silently, and majestically wherever our telescopes point, the galaxies trace out the immense reaches of the Universe, and remind us that our position in it is no more special than that of a boat adrift at sea. Though seemingly immutable, the galaxies too actually do change over the eons of cosmic time. As the nuclear and gravitational energies stored within them are released, galaxies must evolve toward objects ever more structured and compact.

Invisible Revelations

Although every galaxy presents itself as a unique individual, we can discern several recurrent patterns among them – spirals, ellipticals, lenticulars, and irregulars. Ellipticals and lenticulars seem to contain almost no interstellar matter, whereas spirals and irregulars contain enormous masses of interstellar gas and dust, and large numbers of young stars that must have recently formed from them. Until a few years ago, the matter within spiral galaxies other than our own Milky Way could be studied with high spatial resolution only at optical wavelengths. Thus, photographs reveal individual bright groupings of fainter stars and emission nebulae, as well as dark interstellar clouds silhouetted against these bright objects.

Now, the techniques for imaging invisible radiation are rapidly being applied to Nature's grand structures. For instance, the *Very Large Array* radio telescope noted in Chapter 1 can image galaxies using the 21-centimeter radiation produced by their interstellar atomic hydrogen as well as the emission produced by fast-moving electrons gyrating in their interstellar magnetic fields; either technique enables us to map interstellar matter in other galaxies with a spatial resolution comparable with that of optical telescopes.

Somewhat more surprisingly, radio techniques are also helping us discern the halos of galaxies – huge regions of thin gas extending well beyond the hundred-thousand light-year limit considered, until recently, to be the typical dimension of most galaxies. We are beginning to suspect that most galaxies are much larger than their optical photographs suggest. Their invisible gases stretch so far beyond their visible images that they sometimes seem to merge with the invisible extensions of neighboring galaxies.

X-ray techniques are now being used to map the very hot and totally invisible gas that often occupies the space among galaxies, when they cluster together in space. Few galaxies are found isolated in space; virtually everywhere they reside in small groups, and occasionally they form clusters of a thousand or more galaxies. Even the minor groups of galaxies outside the major clusters are not sprinkled randomly through space but rather tend to lie in great sheets and filaments, leaving vast regions apparently devoid of all matter. A question affecting our understanding of the evolution of the Universe is this: Are these regions truly voids, or only apparently so? Perhaps study of invisible radiations from rarefied gas will help answer this question.

Formation of Galaxies

The simplest models of the Universe, developed in the 1920s, are based upon the assumption that the matter in space is distributed with complete uniformity, like the air in the room where you are sitting. This assumption seems to apply well enough to the first 100,000 years in the history of the Universe, when the intense radiation we now detect as the cosmic microwave background permeated all matter and prevented it from clumping together gravitationally, just as the random motions of molecules hinder clumps of air from forming in your room.

But even the simplest models of the Universe show that the matter and its accompanying radiation must cool down as the Universe ages, falling to a temperature of about 4000 degrees some 100,000 years after the big bang. At this temperature the protons and electrons, which up until then comprised the bulk of the matter, were able to combine together to form hydrogen atoms, which interact with radiation much less effectively than had the free electrons before that time. Thereafter the "fog" provided by the free electrons lifted, and radiation was free to travel large distances throughout the Universe. As a result, atoms in space were no longer prevented from clumping together by their mutual gravitation. Calculations indicate that if there were regions where the gas density had been as little as 1/10 of 1 percent greater than the average, before too long these regions would have separated into distinct clouds of gas, each being drawn together gravitationally to form an ever-denser cloud. Could such clouds be the precursors of the millions of galaxies we see today sprinkled throughout space? Perhaps; at least scientists are working on that hypothesis, as we shall see in Chapter 7.

The process we are contemplating is not unlike that of the formation of raindrops in our own atmosphere. Moisture-laden air, warmed by sunlight at the surface of the Earth, rises to higher levels of the atmosphere, which are normally cooler. There the air cools off, and the moisture condenses into droplets of water. In a similar way, the primordial gas emerging from the big bang was cooled by the aging of the Universe to the point that gravitational forces could draw them together to form gas clouds that became galaxies.

Analyzing the situation further, scientists have found that before the critical moment – 100,000 years – when the gas could begin to clump, there are two distinct types of disturbances that could have been present – those in which the temperature is constant from place to place, and those in which it varies in step with the density of gas. The behavior of the two types of disturbances is very different. Variable-temperature disturbances, whose dimensions are such that the denser-than-average regions contained less than a trillion solar masses, would have died out with time, and only those regions larger than a trillion solar masses would have been amplified. The reason for this lies in the behavior of the radiation trapped in the dense regions. Because it exerts pressure, the radiation causes an oscillation in the gas density, just as in a sound wave. But the trapping is not complete if the region is too small, and the radiation diffuses away after a few oscillations, causing the wave to perish. If the region is large enough, the wave survives until the era of gravitational clumping. On the other hand, constant-temperature disturbances embracing very small amounts of matter would have survived, as the pressure of radiation in such disturbances is constant throughout space anyway, and the matter is distributed quite independently.

With powerful computers, researchers have explored the process of gravitational clumping. Thinking that constant-temperature disturbances were more likely, they have expended the largest amount of work on them. As might be expected, small clouds form first and then agglomerate to form galaxies; as time passes, the galaxies themselves agglomerate to form large units such as clusters, sheets, and filaments. Because the results appear to mimic the real Universe, many astronomers have been encouraged by this work to think that galaxies really did form from constant-temperature disturbances in the early Universe.

But that might not be the whole story, for it has been pointed out that in the vast agglomerations of galaxies in space, there is a tendency for elliptical galaxies to be found in regions where the numbers of galaxies are greatest, whereas spirals are found where the numbers of galaxies are low. If the galaxies formed first and clusters formed later, why would the ellipticals be segregated from the spirals? After all, only the gravitational force between any two galaxies is involved, and that does not depend upon whether the galaxies are spiral or elliptical.

Moreover, solution of the problem of galaxy formation demands that we explain why constant-temperature disturbances should dominate the variable-temperature type. Indeed, why were there any disturbances at all? Until recently, scientists had no answers to such questions, which seem to be in the same category as the question, "Where did the Universe come from?" However, in an unanticipated development in the past few years, particle physicists have shown that the conditions in the extremely early Universe, only a fraction of a second after the big bang, might tend to favor the creation of disturbances of the variable-temperature type. The physicists have been able to calculate, tentatively, the dimensions of the denser-than-average regions, and the amounts by which the density should be greater. The results seem to indicate that such disturbances would be of large enough scale to survive – greater than a trillion solar masses – and thus, that the first objects to form may have been far larger than individual galaxies. This would lead to a scenario in which the galaxies formed after the huge clouds of which they were a part were already quite compressed. Such a scenario might open the way to understanding why different types of galaxies would form in different environments, as observed, because the matter was still gaseous in their surroundings, and of different densities.

No one currently knows which if either scenario of galaxy formation is correct. Did constant-temperature disturbances dominate, and lead first to small objects which later accumulated to form galaxies and finally clusters of galaxies? Or did variable-temperature disturbances dominate, leading first to clouds far larger than galaxies, from which galaxies and later smaller units formed? In the years ahead, astronomers will be searching for ways to discriminate between these possibilities.

Evolution of Galaxies

Once galaxies form (by whatever mechanism), they thereafter evolve slowly as supernova explosions inject newly created atomic nuclei into the remaining interstellar gas from which new generations of stars are formed, thus enriching the gas with heavy elements such as iron and magnesium. Our current models suggest that most of the *matter* entering interstellar space is ejected by stars of moderate mass, which evolve into luminous cool stars and then expel puffs of gas seen today as planetary nebulae; as noted earlier in Chapter 3, the luminous cool stars lose most of their outer layers in low-velocity stellar winds, while the deeper layers are thrown off in the final bursts of activity that produce planetary nebulae. Most of the *energy* injected into interstellar space, on the other hand, comes from the explosions of very massive stars, in which entire stars are disrupted to form supernovae. Such catastrophic explosions are also the principal sources of the nuclei of elements heavier than hydrogen and helium, as discussed in the next chapter.

Only stars whose mass equals or exceeds that of our Sun have evolved significantly in the 15-billion-year age of the Universe. Consequently, during the generations of stars that have occured since the galaxies formed, much of the interstellar medium has found its way into low-mass stars that have evolved little over the lifetime of the Universe. We are thus confronted by an essentially irreversible process: interstellar matter is continually enriched in heavy elements, while its mass is continually reduced as more of it converts into slowly evolving stars.

Star formation apparently proceeds at different rates in spiral and elliptical galaxies, since spirals currently contain large amounts of interstellar gas and dust, whereas ellipticals contain little. This is demonstrated by both the correlation of dusty regions with spirals (and not ellipticals) as well as the fact that 21-centimeter radio emission from spiral galaxies is strong, while that from ellipticals is often weak or absent, implying that atomic hydrogen is missing in the latter. We suspect that, early in the life of an elliptical galaxy, the star-formation rate was very high. The massive stars soon exploded as supernovae in such numbers that the ensuing conflagration drove the remaining gas from the galaxy, thus eliminating the material needed for further star formation. We can envision such an outflow of gas as forming a "galactic wind", in analogy to the solar and stellar winds that are directly observed. Supernova explosions occur frequently enough today to keep ellipticals swept clean of interstellar matter. In spirals, on the other hand, the initial supernova rate might not have been great enough to cause a catastrophic purging of interstellar space, and so a sufficient amount of interstellar matter remains today to support active star formation. Thus, the differing amounts of interstellar matter in spirals and ellipticals might result from the different initial rates of formation of massive stars, which later become supernovae. Why the rates of star formation might have differed is an unsolved problem of galactic evolution that can be addressed only by observing galaxies as they were long ago. In principle, because looking out into space is equivalent to looking back into time, this may be possible by studying galaxies having great distances. This will be difficult because of the faintness of the galaxies involved, but x-ray astronomy offers some hope, as we shall see below.

The existence of galactic winds seems to be quite consistent with current x-ray observations of rich clusters of galaxies, especially those clusters in which most of the galaxies are either ellipticals or their close cousins, the lenticulars. In many such clusters, we find hot, x-ray-emitting intergalactic gas whose total mass and chemical composition are consistent with accumulations of galactic winds from the various member galaxies of the cluster. (The intergalactic medium in a cluster of galaxies is gas outside individual galaxies but still within the cluster. Genuine intergalactic gas outside of clusters has not yet been identified conclusively by astronomers, although there are tantalizing hints that even the vast regions between clusters of galaxies – the voids mentioned previously – are not completely empty.) Removal of loose gas from galaxies is further aided by any gas already in intergalactic space because it can sweep gas from galaxies as they move through it. Recent radio and x-ray images show dramatic evidence that galaxies are being swept clean of any stray interstellar matter not bound in stars.

A further advance in studies of galaxy evolution is the recent recognition that galaxies interact with one another as well as with the intergalactic gas around them. In fact, galaxy collisions occur fairly often, especially in rich clusters. During such a collision (or even a close encounter), the evolution of a galaxy is likely to be modified in a variety of ways as mass, energy, or angular momentum are exchanged with the other galaxy. Take the case of spiral galaxies, which we now realize have huge, invisible halos where substantial mass is hidden in faint stars or in some other invisible form of matter. Such galaxies can orbit around one another, forming a "binary galaxy". As they do so, they tend to interact with each other's halos, one galaxy stripping the halo material from the other by tidal forces; the freed matter is then either redistributed within a common envelope or is entirely lost from the binary system. Since any energy or angular momentum lost in this way must come from the motion of the galaxies in their binary orbit, the orbit itself is forced to change its size and/or shape. Moreover, if one galaxy of the pair has a low mass, it may end up spiralling into the other, finally merging with the more massive galaxy; since such events invariably make the more massive galaxy still more massive, the process is colloquially termed "galactic cannibalism" or even "galaxy gobbling". Such cannibalism might explain the fact that supermassive galaxies are often found at the cores of rich galaxy clusters. Having dined on their companions, they are now content to relax at the center.

One final point on this subject: Any realistic scenario for galaxy evolution must take into account the effects of cosmic rays – those fast-moving particles described in Chapter 2 whose presence we infer from experiments with charged-particle detectors on Earth-orbiting satellites. Comprising a prominent and permanent component of our Galaxy, the relative abundances of the nuclei of elements in cosmic rays imply that they remain in the Galaxy for tens of millions of years, much of that time in an extended halo beyond the flat disk where the spiral arms are found. In effect, cosmic rays comprise a gas whose pressure on any surface is comparable to that exerted by the random motions of the interstellar gas. Accordingly, cosmic rays play a critical role in trying to push interstellar gas up out of the disk. Determining the origin and understanding the propagation of cosmic rays is thus an important aspect of our overall effort to decipher the course of galaxy evolution, and we shall have more to say about them in the next chapter.

Future Prospects

As in all fields of astronomy, spectroscopy is the key to deeper understanding. Ground-based optical spectroscopy of galaxies has demonstrated that stars of various masses and ages, much like those in our own Milky Way, comprise a major component of most galaxies. However, current ground-based telescopes are hard pressed to obtain the spectra of extremely faint subsystems within galaxies, such as individual luminous stars, emission nebulae, and star clusters; these systems are too small and their parent galaxies too far away for us to collect radiation of sufficient intensity. The *New Technology Telescope* (*cf.*, Appendix F), with its order-of-magnitude increase in collecting area, could obtain the spectra of such objects, thus facilitating a whole new area of studies concerning the chemical composition of stars, the range of stellar masses, and the velocities of objects within galaxies.

To address the large-scale distribution and clustering of galaxies in space, and especially to penetrate more deeply into space, we shall also need larger telescopes. Optical instruments such as the currently available 5-meter class telescopes will continue to make valuable contributions, but only a new telescope of the 15-meter class, such as the *New Technology Telescope,* can measure the Doppler shifts of galaxies at remote distances to accumulate data rapidly enough to unravel the underlying structure of clusters, superclusters, and voids in deep space. The raw speed of this newly proposed "monster" telescope is critical for this project.

One of the most striking capabilities of the new instruments likely to be built during the remaining years of this century is the systematic exploration of the dependence of various galaxy properties on distance from us. Big-bang models of the Universe predict such a dependence because galaxies are expected to evolve over the course of time, and this translates into changing galaxy properties as we look farther into deep space. Both the *Space Telescope* (*cf.*, Appendix A) and the *New Technology Telescope* will be able for the first time to observe galaxies halfway to the limits of the observable Universe. These two new telescopes, one in orbit and the other on the ground, will work together; the *Space Telescope* should be able to readily discover distant objects because the sharpness of its images makes them stand out against the background, while the *New Technology Telescope* can obtain their spectra in a reasonable time because it will have a much larger collecting area than any other current or foreseeable telescope.

What sort of evolutionary changes can we expect to perceive in galaxies? If the cores and innermost parts of galaxies form, as expected, within the first few billion years in the history of the Universe, then even these new instruments will not likely address how the overall forms of galaxies change; even the most distant galaxies they can study are not young enough to differ from nearby galaxies as regards their overall structures. However, the evolution of stars and the conversion of interstellar gas into stars proceed in a much more drawn-out fashion, and we should be able to discern, for example, subtle changes in the spectra of galaxies that reflect the evolution of stars in elliptical galaxies; likewise, in spiral galaxies we should be able to detect, in addition, the progressive depletion of interstellar matter, as well as its enrichment in heavy elements produced by supernova explosions. A major indirect effect expected is a reduction in the number of short-lived massive stars as the gas required to form them is depleted. Failure to observe such basic predictions of the big-bang theory would force major revisions in our current thinking.

As for exploring the early times when galaxies presumably did form, two major instruments now in the design and development stage may yield significant information. By observing the disturbances in the cosmic background microwave radiation flooding the Universe, the *Cosmic Background Explorer* satellite, scheduled to be orbited in the mid-1980s, will yield important information on the suspected gravitational instability processes that led to galaxies. Variable-temperature disturbances involve changes in the intensity of the background radiation on angular scales of a few degrees if the masses involved roughly equal those of clusters of galaxies. Complementary data about disturbances on the smaller angular scales corresponding to individual galaxies (less than an arc degree across) will be obtained in space by the *Large Deployable Reflector* (*cf.*, Appendix G).

As noted above, the contraction of a gas cloud to form a spiral galaxy such as our Milky Way, as well as the subsequent formation of the first generation of stars, are both thought to have occurred during the first billion years of galaxy evolution. These processes can therefore be observed only in galaxies so remote that the radiation now received from them was emitted in their early youth. Although presumably very faint owing to the great distance involved, this radiation might be detectable by the next generation of telescopes. For example, optical and ultraviolet radiation from young galaxies, if not absorbed by dust near the galaxy or in intergalactic space, would be Doppler shifted by the expansion of the Universe (*cf.*, Chapter 7) into the infrared part of the spectrum, where the high sensitivity of the *Shuttle Infrared Telescope Facility* (*cf.*, Appendix C) would likely permit detection. Overlapping shock waves from early supernovae are expected to heat the interstellar gas in galaxies to temperatures as high as a billion degrees, generating dust-penetrating x-rays which should be Doppler shifted into the range of detectability of equipment aboard the *Advanced X-Ray Astrophysics Facility* (*cf.*, Appendix D). Should a galaxy like ours that had just been born be discovered by any one of these newly proposed instruments, it would present a breathtaking opportunity for study.

Verifying our ideas of galaxy evolution will require intensive studies of our own and nearby galaxies, as well as observations of galaxies so remote that their properties appear different from those of the more evolved galaxies nearby. As noted in Chapter 3, we now realize that stars normally lose large quantities of gas, so this mass loss must have profound effects on galaxy evolution. With its high sensitivity, the *Shuttle Infrared Telescope Facility* will discover many more envelopes around luminous cool stars, in which stellar light is degraded to infrared radiation by large amounts of embedded dust; several new telescopes, including the *Large Deployable Reflector* in space, the *Submillimeter-Wave Radio Telescope,* and the *New Technology Telescope,* will be well suited to make spectroscopic observations of molecules in such envelopes, leading to determinations of chemical abundances, which are clues to the nuclear processing occurring in the parent stars. These instruments will also be able to measure the velocity and amount of the outflowing gas, crucial parameters for calculating the rate at which mass is being ejected into interstellar space.

Because of its ability to study exceptionally faint objects, the *Space Telescope* will be able to identify a much larger fraction of the low-mass stars suspected to reside nearby in our part of the Milky Way, thus permitting a much more accurate assessment of the mass stored in these stars; these studies will be aided by the *Shuttle Infrared Telescope Facility,* which will be much more sensitive to the lukewarm faint stars than are the infrared telescopes now available. Because of its ultraviolet capabilities, the *Space Telescope* is expected to find many new white dwarfs – burned-out but very hot remnants of ordinary stars – hence improving our estimates of the number of moderate-mass stars that have already experienced much evolution. The *New Technology Telescope,* whose large collection area will make it possible to obtain spectra of various subsystems of nearby galaxies, will reveal variations in abundances expected to develop because of different rates of evolution at different points within galaxies.

Continued on page 95

(above) A part of our Milky Way galaxy as seen with the Infrared Astronomical Satellite (IRAS). The field of view is approximately 48 × 33 degrees. The large yellow area at the center of the colored band is the Galactic Center. The yellow and green knots scattered along the band are giant clouds of interstellar gas and dust that glow from the emissions of nearby stars. This image is a composite of multiple wavelength bands: 100 microns (red), 60 microns (green), and 12 microns (blue). The extended blue regions are emissions from the dust in the solar system, which has been warmed by the Sun.

(below) The region of the sky away from galactic center as observed from the Infrared Astronomical Satellite. This image, shown in an Aitoff projection, shows a survey of the entire sky as observed by IRAS during its first six months of operation. The image coloration corresponds to the various wavelength bands observed: 100 microns (red), 60 microns (green), and 12 microns (blue). The dark arcs are regions of the sky that were not scanned. The blue S-shaped figure is produced by radiation from warm dust in our solar system.

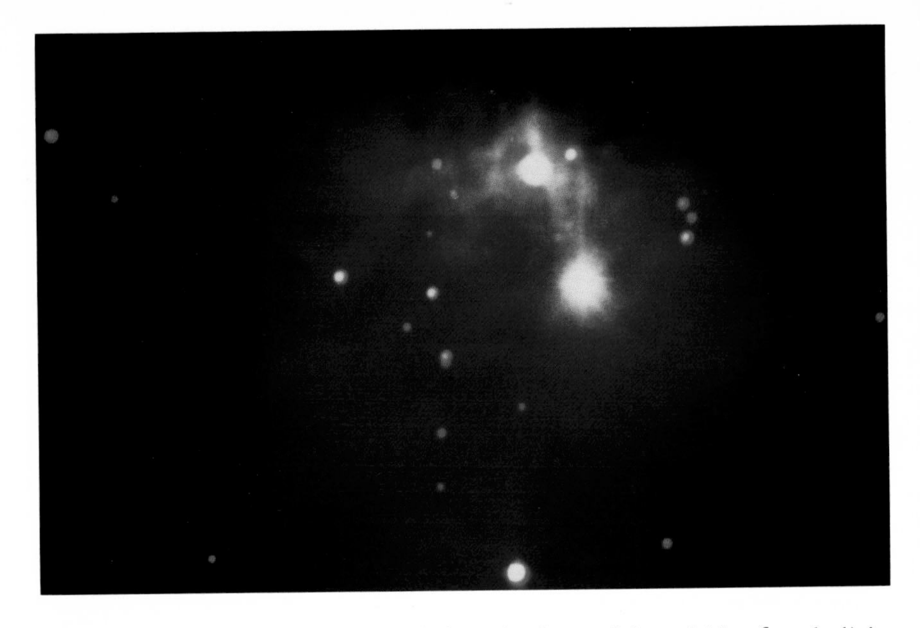

(above) NGC 2068, a region composed of clouds of gas and dust which reflect the light of two young stars that formed in the last million years. The upper star is redder because its light passes through heavier clouds of absorbing material. A vertical row of nine stars to the left of center apparently formed recently where two large clouds of gas and dust collided.

Optical image recorded with a CCD camera on Polaroid Polacolor ER Land Film Type 809.

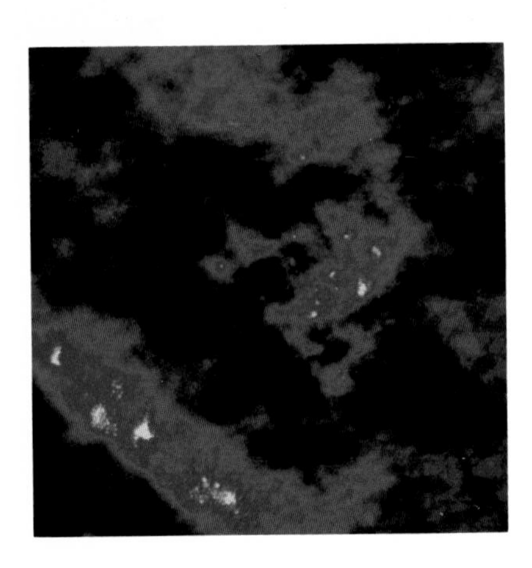

(left) The Orion Nebula (M42) recorded by the array of radio telescopes in Socorro, New Mexico, known as the Very Large Array. The nebula is composed of interstellar gas and dust. At visible wavelengths, light from nearby stars excites the atoms in the nebula and causes them to glow. However, at radio wavelengths the emissions originate from the hot gasses themselves. The radio image shows many of the same features recorded in the familiar optical image of Orion. In particular, note the "bar" in the lower left and the bright area in the upper right known as the Trapezium.

Radio image from the VLA recorded on Polaroid Polacolor ER Land Film Type 809.

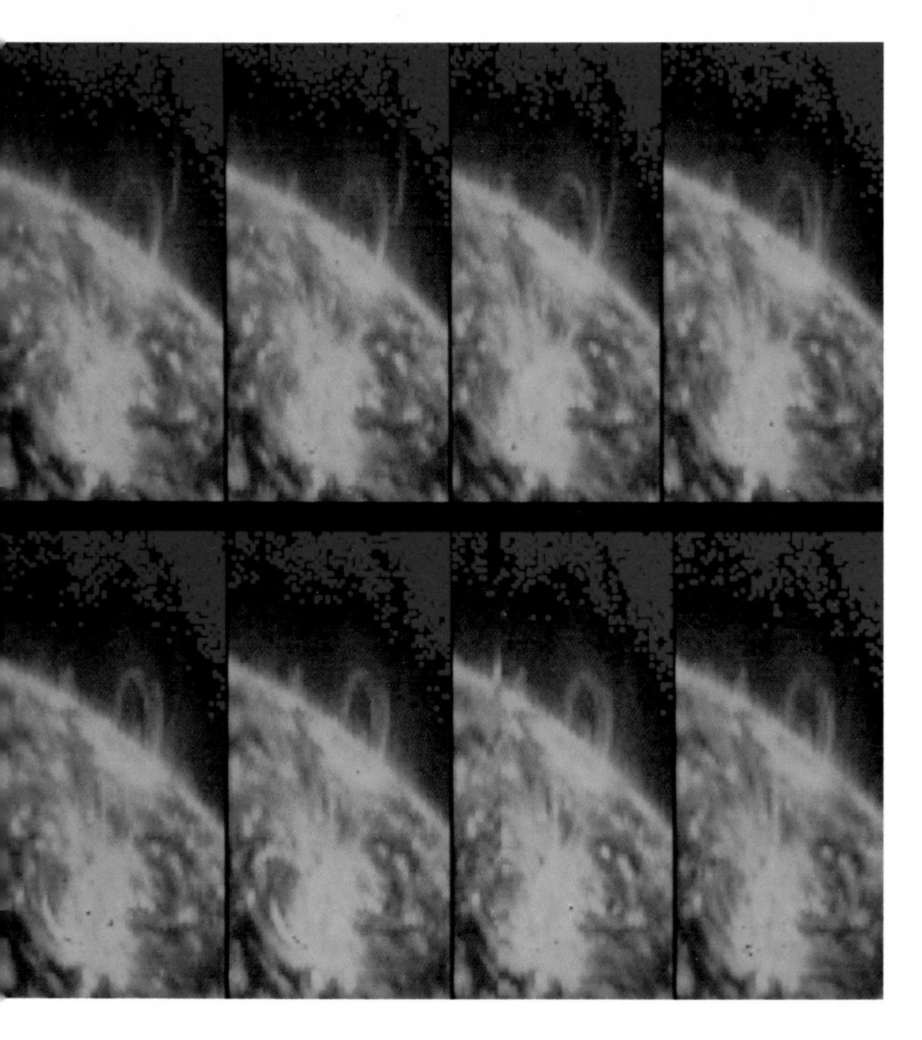

Coronal loops near the edge of the Sun. The Sun's magnetic field causes the hot, ionized gasses above its surface to form loops or arches as shown here in a time sequence. This sequence of images extending from left to right shows several developing loops spanning a time period of approximately 5.5 minutes. These images were made using the emission from neon gas in the extreme ultraviolet wavelength range; they were made by equipment on board Skylab, a manned, orbiting space station equipped with an observatory to permit solar studies at wavelengths not visible from the Earth.

Ultraviolet image recorded on Polaroid Polacolor ER Land Film Type 809.

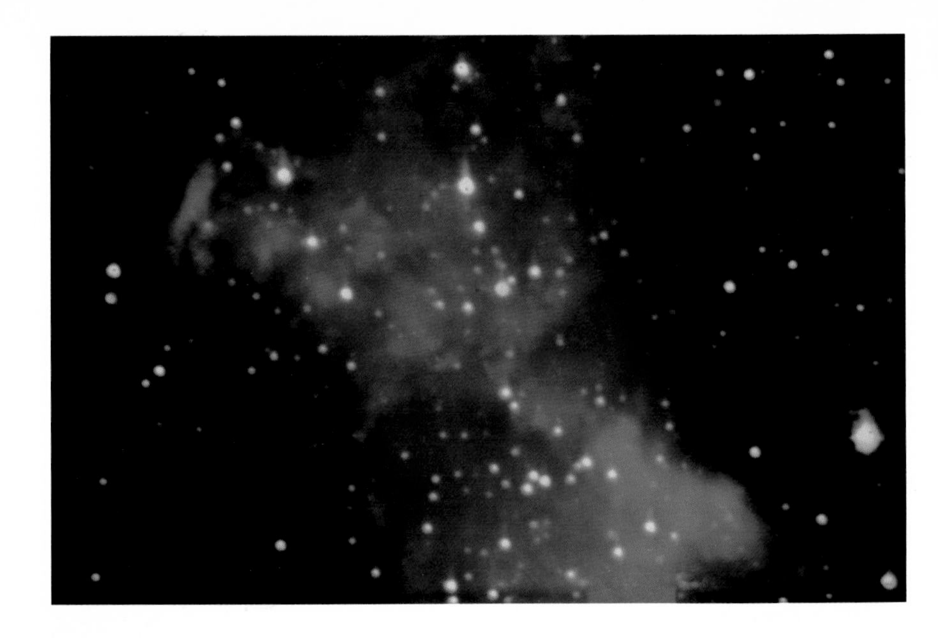

(above) The Dumbbell Nebula is a planetary nebula, a beautiful gas cloud that was ejected from a dying star. It will continue to expand and become more diffuse. Its colors result principally from primary constituents of the gas cloud, hydrogen and oxygen. As the cloud expands, it will gradually merge with the interstellar gas and will become part of the next generation of stars. This image was captured with a CCD camera using three filters through which pass a unique set of wavelengths. These have been recombined in the image processing system to reveal the enhanced color photograph shown here.

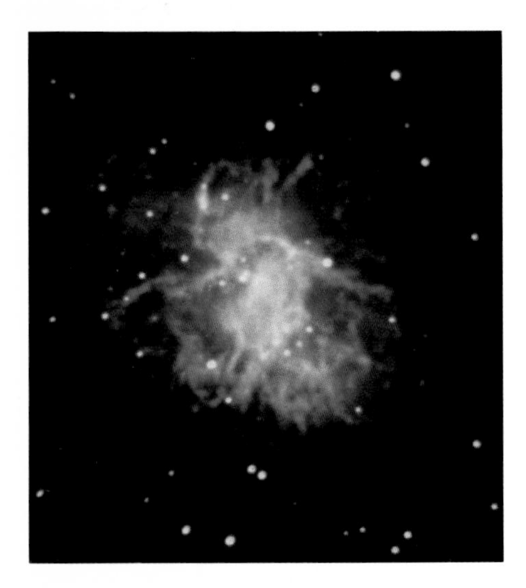

(left) A CCD image of the Crab Nebula, shown here in visible and infrared wavelengths. The Crab is the result of a supernova explosion first seen by Arab, American Indian and Chinese observers in 1054 A.D. The brilliant red filaments indicate the presence of ionized hydrogen, while the diffuse blue-white glow, known as the "synchrotron" nebula, is due to radiation emitted by fast-moving electrons trapped in the nebula's magnetic fields.

Optical images recorded with a CCD camera on Polaroid Polacolor ER Land Film Type 809.

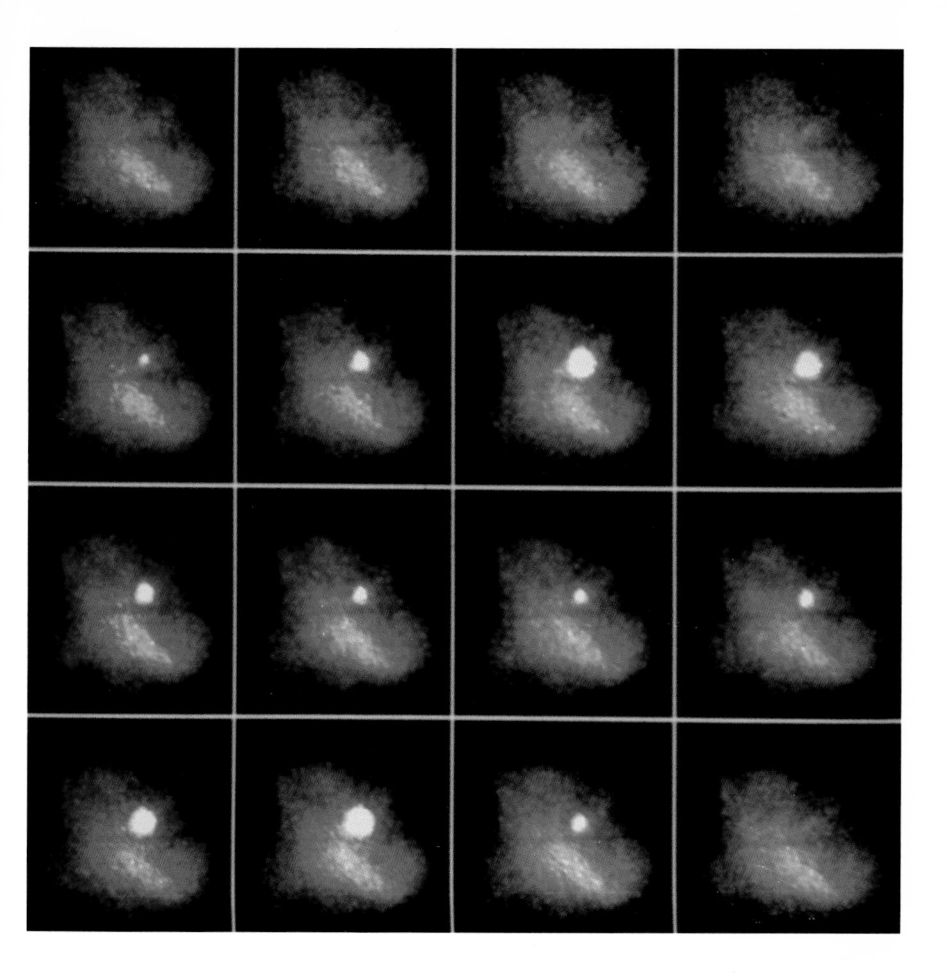

X-Ray emissions of the Crab Nebula as recorded in a time sequence by NASA's Einstein X-Ray Observatory. The bright spot in the center of the nebula is the now-famous Crab pulsar, which emits pulses about 30 times per second. These pulsations, first discovered by radio astronomers in 1968, have since been observed both at optical and x-ray wavelengths. The pulsations are most probably due to intense beams of radiation emanating from a rapidly spinning neutron star at the center of nebula, a remnant of the original supernova. Note the successive brightening and fading of the pulsar as time proceeds.

X-Ray image from Einstein X-Ray Observatory recorded on Polaroid Polacolor ER Land Film Type 809.

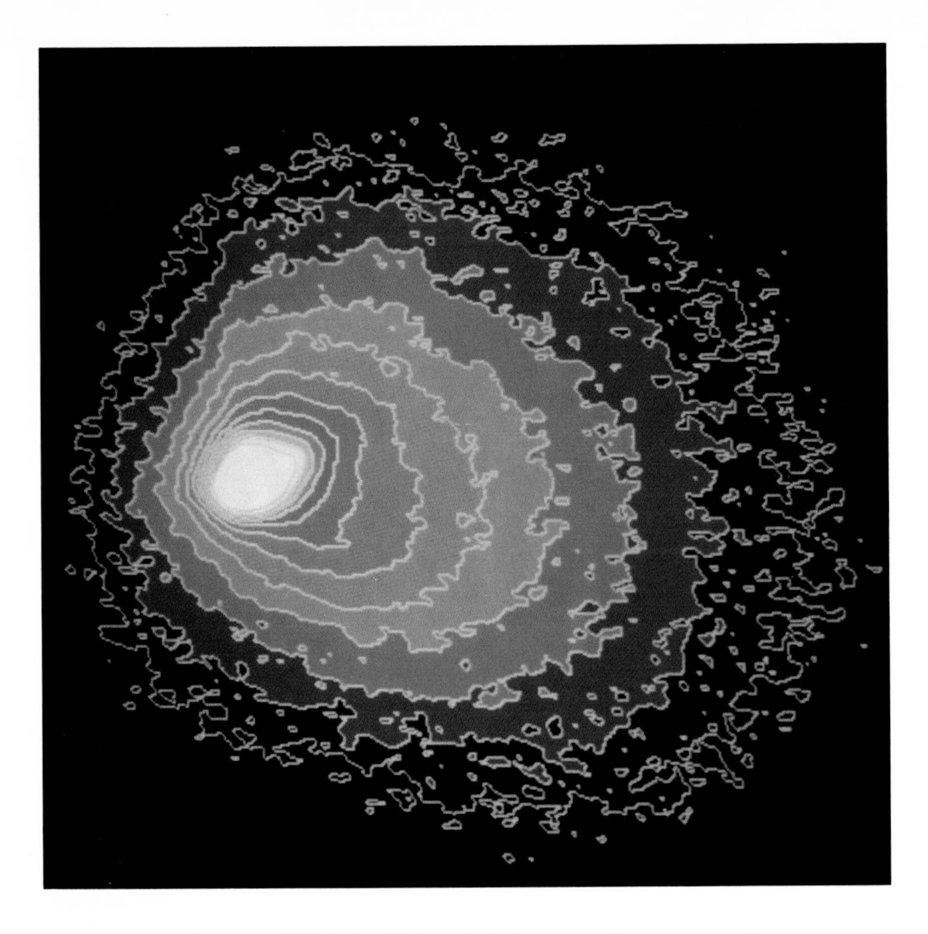

Comet 1983d, known as IRAS-Araki-Alcock for its three discoverers, as first seen on 11 May 1983. As the comet tumbles in space, it presents a continuously changing face to the sun. This causes variations in the amount of gas and debris that streams away from the side facing the sun. Here we see the comet presented with distinct intensity contours. Note especially the changes in the shape and orientation of the contours at different distances from the comet, indicating variations in the amount of ejected material.

Optical image recorded with a CCD camera on Polaroid Polacolor ER Land Film Type 809.

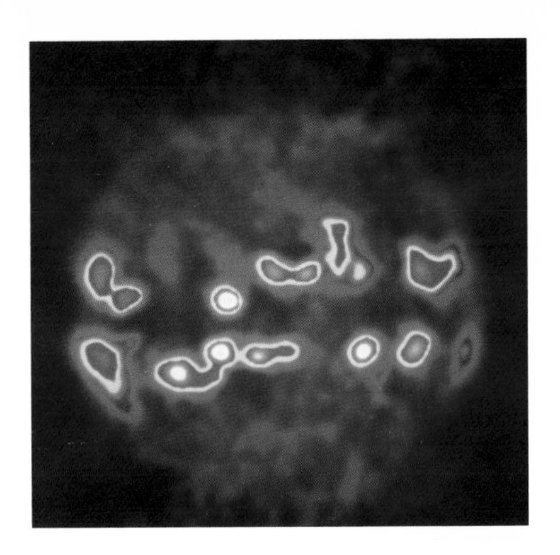

(above) The Sun as viewed at radio wavelengths using the Very Large Array (VLA). Especially noteworthy are the active solar regions that form two symmetric bands about the equator. Each active region is in the vicinity of, but situated above, corresponding sunspots. The relatively dense, hot material in active regions is confined by strong magnetic fields emanating from sunspots.

(below) Saturn as observed at radio wavelengths with the VLA. The bright disk of the planet fades towards the edge, an effect caused by the fact that the upper atmosphere is cooler and hence emits less radiation than the lower atmosphere. Saturn's rings are seen as emitters of radio energy in the area outside of the disk, but as absorbers of energy where they eclipse the planet's surface. This is in contrast to the situation at visible wavelengths where the rings reflect the incident sunlight everywhere.

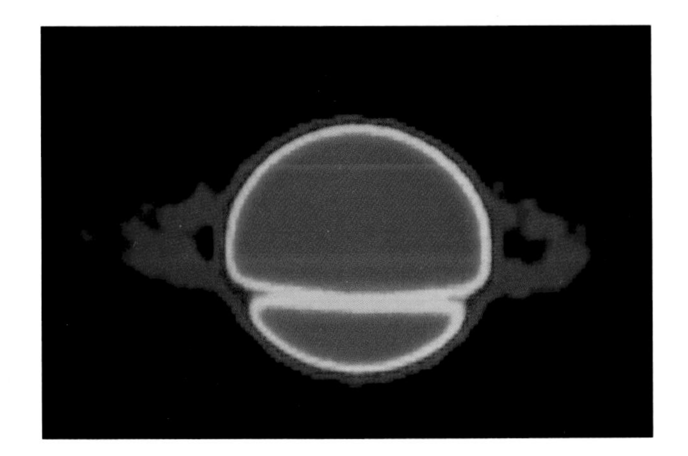

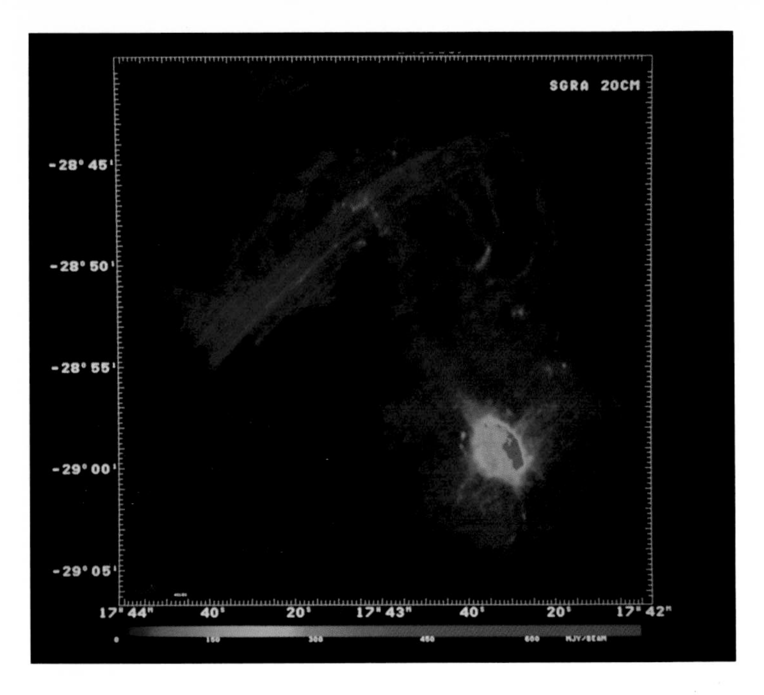

(above) The center of our galaxy, the Milky Way, is shown here at radio wavelengths in a composite image made by the VLA. The bright ring of material in the lower left is due to a radio source known as Sagittarius A, which contains the galactic nucleus. The plane of the galaxy is parallel to the long axis of this ring and intersects the striking filamentary radio structures (shown in blue) at a right angle. The filaments have been interpreted as due to explosive activity in the nucleus.

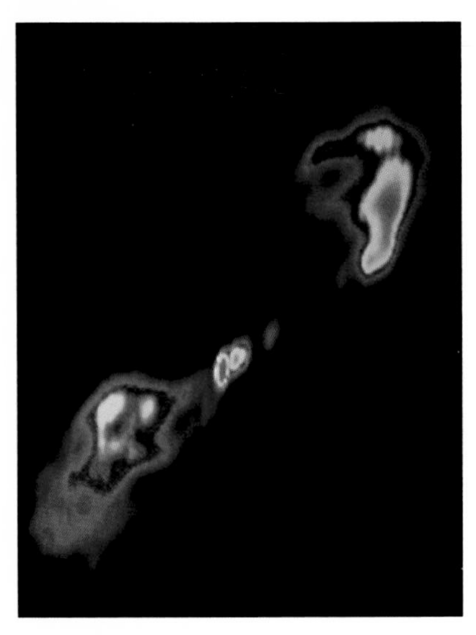

(left) The giant galaxy NGC 5128, located about 16 million light years from Earth. This galaxy has three times as many stars as the Milky Way and is visible at many wavelengths from radio to x-ray. NGC 5128 consists of both very old and very young stars, leading to speculation that it is really two galaxies, an elliptical galaxy that has captured a spiral galaxy. Shown here are the inner lobes of the radio source Centaurus A associated with NGC 5128 as observed with the VLA. The galactic nucleus is the larger of the two red spots at the center of the image. The lobes of radio emission shown here are about 30000 light years across. The radio jet associated with this galaxy shows a strong correlation with x-ray observations made by the Einstein X-Ray Observatory.

By yielding sharp images of very distant galaxies, the *Space Telescope* will permit them to be sorted into ellipticals, spirals, lenticulars, and irregulars for the first time; this is essential in order to interpret correctly the observed correlations between distance and other properties of galaxies. In recent years, there have been exciting hints that remote galaxies tend to be systematically bluer than their nearby analogs, but, because classifying these galaxies is beyond the capability of current ground-based telescopes, we are unsure how to interpret these observations. The *New Technology Telescope* will also play an important role in this research by permitting us to obtain optical and infrared spectra of the faraway galaxies, especially those new ones discovered by the *Space Telescope.*

Regarding the interactions of galaxies – with one another as well as with their intergalactic environments – the gravitational fields of such systems will be probed via measurements of the velocities of stars within them using the *New Technology Telescope,* as well as by studies of the distribution of hot gas around galaxies with the *Advanced X-Ray Astrophysics Facility.* Evidence from x-ray observations of the Virgo galaxy cluster some 50 million light-years away suggests that its centermost member, the giant elliptical galaxy Messier 87, is currently accreting large amounts of intergalactic gas. As we discuss more thoroughly in the next chapter, a leading theoretical model stipulates that the intense emission from this galaxy arises from the production of energy by the flow of matter into a supermassive black hole located in the galaxy's core. Such accretion of intergalactic gas in galaxy clusters might provide a virtually unlimited source of matter to power the active galaxies often found near the cores of galaxy clusters. Here, the *Advanced X-Ray Astrophysics Facility,* with its ability to observe the diffuse, hot gas with high spatial resolution and increased sensitivity will permit us to study many active galaxies at larger distances for evidence of the M87 phenomenon.

The whole problem of activity in the centers of galaxies poses a major puzzle in our studies of galaxy evolution. Both elliptical and spiral galaxies occasionally display activity that is highly localized in their central regions. While we discuss some possible physical explanations for such activity in the next chapter, we should note here those aspects of galaxy activity that are relevant to galaxy evolution generally. In particular, we have the vexing issue represented by the key question: Do those extreme examples of cosmic activity we call quasars occur within galaxies? Or are the quasars isolated explosions in space? The *Space Telescope,* with its fine spatial resolution, should be able to detect the parent galaxies, provided they exist; the light from a parent galaxy, which is lost in the glare of the blazing quasar when studied with present ground-based telescopes, will likely be distinct from the tiny quasar image when viewed with the *Space Telescope.* Do many quasars occur within groups of galaxies, as recently implied by ground-based observations? Again, the *Space Telescope,* because its superb sensitivity will enable us to image the accompanying normal galaxies, and the *New Technology Telescope,* because of its ability to acquire spectra of faint objects, will be able to address this question even for distant quasars, thus determining whether the occurrence of a quasar in a galaxy depends on the environment of the galaxy. Is the density of stars in active galaxy centers high enough to explain such activity by stellar collisions? High-resolution images made with the *Space Telescope* will penetrate close to the centers of some galaxies and thereby help answer this question directly; its spatial resolution of 0.05 arc second corresponds to about 100 light-years at the distance of the nearest active galaxies. Is there an inward flow of hot gas in active ellipticals, as required by theories based on accretion by a massive black hole? The high-resolution images scheduled to be obtained by the *Advanced X-Ray Astrophysics Facility,* and especially the studies of the time variability of emission that this instrument will be able to perform, might help to answer this question as well.

The astronomical equipment scheduled to be built during the remaining years of the twentieth century will have a major impact on studies of galaxy evolution. Over the entire range of phenomena – from the earliest development of density fluctuations in the primordial Universe, through star formation in contracting young galaxies, to the slow conversion of interstellar gas into stars, and the emergence of activity in the centers of galaxies – observations by the new instruments in the radio, infrared, optical, ultraviolet, x-ray, and gamma-ray regions of the electromagnetic spectrum are destined to provide a wealth of new information.

Extragalactic Water

Having arrived at Harvard as a new graduate student just about the time (1970) when the first interstellar molecules were being discovered, I knew that it might be exciting – and significant – to extend those discoveries to other galaxies. I enlisted the help of Smithsonian radio astronomer Dale Dickinson, and together we went molecule hunting in some of the nearby galaxies akin to our Milky Way. Using the Haystack telescope in Massachusetts, we concentrated on finding the water molecule, firstly because its natural maser emission in our own Galaxy makes the H_2O signal very strong, and secondly because water is a key ingredient for life as we know it. After many attempts, especially toward the giant spiral system M33, we were unable to find any evidence of water. Off and on for the next several years, we searched to no avail for the essence of life in other galaxies.

By the mid-1970s, the Haystack telescope had been outfitted with much more sensitive equipment, and we tried the experiment again. This time we succeeded! We were elated to know that water vapor existed in huge quantities far beyond our Galaxy. I remember celebrating our discovery with a drink – in this case, appropriately of water that I had secured from the observatory kitchen. But, before our finding could be announced, we knew that we had to confirm it in order to be absolutely sure of its validity, for reproducability is a hallmark of observational science.

As I tooled up to prepare to confirm our finding on the second night of our observing run, I took a break to watch the evening news. And then it happened: Walter Cronkite announced that a group of German radio astronomers had discovered, a few days earlier, the existence of water in another galaxy. They had found evidence of the molecule at virtually the same site in M33 as we had on the previous night, and they had already confirmed their findings. We had lost our multiyear search for water by just a few days.

This is a mild example of competition in science. At any one time, the next key experiments to be done are usually obvious to those working at the frontiers, and several groups often vie for the discoveries. Only one group ever wins, as there are no silver and bronze medals in science. Even so, regardless of whose ego is inflated by being first, knowledge advances and we all learn.

E.J.C.

6
Cosmic Violence
Black Holes, Quasars, and Beyond

How, in order that we may harvest some fruit from the unexpected marvels that have remained hidden until this age of ours, it will be well if in the future we once again lend ear to those wise philosophers whose opinion of the celestial substance differed from Aristotle's. He himself would not have departed so far from their view if his knowledge had included our present sensory evidence, since he not only admitted manifest experience among the ways of forming conclusions about physical problems, but even gave it first place. So when he argued the immutability of the heavens from the fact that no alteration had been seen in them during all the ages, it may be believed that had his eyes shown him what is now evident to us, he would have adopted the very opinion to which we are led by these remarkable discoveries. I should even think that in making the celestial material alterable, I contradict the doctrine of Aristotle much less than do those people who still want to keep the sky inalterable; for I am sure that he never took its inalterability to be as certain as the fact that all human reasoning must be placed second to direct experience. Hence they will philosophize better who give assent to propositions that depend upon manifest observations, than they who persist in opinions repugnant to the senses and supported only by probable reasons.

From a letter to one of his patrons,
by Galileo Galilei, Florence, 1612

PLANETS, STARS, AND GALAXIES thus far noted here represent a truly varied lot, displaying not only the quiescence recognized by our ancestors but also a rich spectrum of activity much more recently discovered. For example, we have discussed the solar and stellar activity traceable to magnetic fields. Sometimes, however, the activity of stars and galaxies becomes violent, threatening the very existence of the object under study. In this chapter, we discuss some of the truly violent objects in the Universe. Violent objects are those most likely to challenge the known fabric of science; for that very reason, they are also the regions where we might discover new laws of nature. Cosmic violence, like cosmic activity and quiescence, demands an explanation.

Cosmic Rays, Supernovae, and Bizarre Stars

Earth is constantly bombarded by cosmic rays, which despite the name are subatomic particles such as protons and helium nuclei that cruise through space at very high speeds – speeds approaching that of light and therefore termed relativistic. Discovered early in this century, cosmic rays provided the first hints that the Universe is not a quiescent collection of stars and planets, but, rather, the scene of violent events in which particles are accelerated to relativistic energies. In the early years, researchers had to be content analyzing whatever cosmic-ray particles just happened to be intercepted by Earth. Since the cosmic rays are electrically charged and therefore affected by magnetic fields, the bending of their paths by magnetic fields in interstellar space precludes identifying their sources (as we do for radiation) by observing the direction from which they arrive at Earth.

We can divide the mystery of the cosmic rays into two parts: How do they propagate from their sources to us? And where are they accelerated to their high energies? The first question is answered largely by the interstellar magnetic fields (whose presence is known from their effects on observed cosmic radio waves.) In the process of bending the paths of cosmic rays, magnetism confines them for long periods within the Galaxy, where they bounce around until approximately equal numbers of such particles exist everywhere in our Milky Way. This idea is confirmed by gamma-ray astronomy, which detects the gamma rays emitted when a cosmic ray collides with a hydrogen or helium nucleus in interstellar space. If cosmic rays really are distributed evenly in the Galaxy, the gamma-ray intensity should be proportional to the amount of interstellar matter – and that is exactly what is observed by the *COS-B Gamma-Ray Satellite* operated by the European Space Agency.

Where, then, are cosmic rays accelerated? The answer is currently unknown, though we do have some definite clues. Supernova explosions, in which nearly an entire star is disrupted and flung into space at speeds up to 10,000 kilometers per second, seem likely sites, so it is interesting that the remnants of such explosions are among the most powerful radio sources in our Galaxy. In the 1950s, radio astronomers proved that the emission of radiation from supernova remnants arises from the fast movements of large numbers of electrons trapped in magnetic fields. Hence at least electrons are accelerated to relativistic energies in supernova explosions or their immediate aftermath. Applying the same idea to a diffuse galactic background of radio emission, theorists calculated the number of relativistic electrons that must be spread throughout the Galaxy to account for the observed emission. Although such fast-moving electrons had not been detected near the Earth before, there was now good reason to search for them. Subsequently found, the relativistic electrons are present in just the numbers predicted, a result that also proves that such electrons, like the cosmic-ray protons and helium nuclei, are spread rather evenly across the Galaxy. And since at least some relativistic electrons are accelerated in supernova explosions, we can reasonably suppose that some of the cosmic-ray particles are too.

What is the mechanism by which supernovae accelerate particles to high energies? One such mechanism is shock waves. In the vicinity of a supernova explosion, astronomers often observe a giant spherical region that has been overtaken by the shock wave from the explosion. It is quite possible that this shock wave accelerates particles, because, as we shall see in a moment, we actually observe this very process occurring in the vicinity of the Earth. Nowadays, theorists feel that the most likely means of accelerating galactic cosmic rays results from the occasional buffeting of any region of space by a passing supernova shock wave, which simply reaccelerates any energetic particles that happen to be nearby. Such a theory reasonably accounts for the total numbers of cosmic-ray particles, as well as their distribution among various energies.

We now know that supernova remnants are not the only places where electrons and cosmic-ray particles can be accelerated to relativistic energies. A wide variety of astronomical systems do so, from the Sun and planets to the giant galaxies. In fact, particle acceleration seems to occur wherever magnetic fields are embedded in the ionized gases (or plasma) pervading interplanetary, interstellar, or intergalactic space.

Close to home, Earth's magnetic field and its embedded plasma, and especially its Van Allen Belts of magnetically trapped energetic particles, provide opportunities for *in situ* studies of the acceleration process. By directly observing processes in this nearby plasma, our Space-Age satellites have shown that charged particles can be accelerated by the passage of shock waves and the reconnection of magnetic fields. Here, we refer to the same reconnection mechanism outlined earlier for the Sun in Chapter 3; magnetic fields occasionally become unstable and transform into new configurations, while using their energy to accelerate large numbers of particles in a short time. The resulting fast particles probably trigger a wide spectrum of activity, including earthly auroras and solar flares. Moreover, recent spacecraft discoveries of intense flaring on other stars imply that magnetic activity is commonplace, with the flares on some stars being far more energetic than those on our Sun.

Still, particle acceleration near stars and planets is miniscule compared to the more violent events marking the deaths of massive stars. Optical, radio, and x-ray observations of supernova remnants strongly suggest that cosmic rays can be accelerated by supernova shock waves.

Observations have demonstrated at least two distinct types of supernovae. Type I supernovae often occur in binary systems containing low-mass stars, and are discussed in the next section. On the other hand, type II supernovae occur throughout the disks of spiral galaxies, and seem to be a natural step in the evolution of the cores of isolated massive stars. As a star more massive than our Sun evolves, its core undergoes successive stages of contraction and heating while the atomic nuclei within it combine to form increasingly heavy elements. Eventually, a core of at least 1.4 solar masses of pure iron accumulates, at which point, since the fusion of iron to form even heavier nuclei does not release energy but requires it, the core becomes catastrophically unstable and begins rapidly collapsing; at this point, the star's internal pressure drops because any free electrons present are captured by simple atomic nuclei, while energy is being drained off by the disintegration of iron nuclei into elementary particles. If the resultant core is very much larger than 1.4 solar masses, no force develops that is strong enough to reverse its inward momentum, and the matter collapses toward a point, forming what is known as a stellar black hole. This configuration is one of the most perplexing states of matter imaginable. Its gravitational attraction is so great that even light, once within a small distance measured in kilometers, cannot leave it, but spirals in to the point-like mass at the center. Even passing photons are trapped, thus darkening the region from which this bizarre "star" derives its name.

If the massive star has a core of lower mass (although still at least 1.4 solar masses), the collapse might be halted by the repulsive nature of the strong force that exists between atomic nuclei. Such a star's central density approaches and possibly even surpasses that of typical atomic nuclei (*viz.*, $\sim 10^{15}$ grams per cubic centimeter), but we are uncertain what happens next. One possibility is that the collapsing core responds to the repulsive nuclear force by rebounding, thus sending a shock wave propagating into the loosely attached envelope of the star, where it heats the matter to such high temperatures that a brief surge of reactions occurs among the nuclei present. In this way, elements even heavier than iron are created, although only in relatively small amounts, for the expelled matter thins and cools so rapidly as to preclude further nuclear reactions. Significantly, the shock wave not only helps create the elements, but it also ejects those elements into the surrounding interstellar space where they are fated to be collected into new stars. Thus, the deathrattles of intermediate-mass stars are crucial phenomena in the life of a galaxy, for they account for all the elements heavier than oxygen present in the interstellar medium, and hence in later generations of stars, planets, and apparently life itself.

As we might expect, the process of core collapse is a violent event, suddenly releasing in a few seconds an amount of energy which, according to Einstein's formula, is equivalent to the instantaneous complete annihilation of about a tenth of our Sun's mass (or alternatively about 30,000 Earths). Much visible light is emitted from the shell of ejected matter, which expands outward at velocities up to 10,000 kilometers per second. But most of the energy is released invisibly. It may emerge immediately in the form of neutrinos and/or gravitational radiation, (*cf.*, Chapter 8), or it is released much later, when the supernova shell plows into the surrounding interstellar gas, sending into it a shock wave that heats the gas to x-ray emitting temperatures.

Some supernovae leave behind a compact remnant of a few solar masses made exclusively of neutrons. Because such burnt-out cores (called "neutron stars") experience extreme compression before and during the explosion, they are expected to be hardly more than a few tens of kilometers across – much smaller than a planet, and rather more like the dimensions of a city. Packing such a huge mass within a small volume means that the density of a neutron star must be extraordinary, comparable to that of typical atomic nuclei. (To call upon comparisons with more familiar phenomena, the force of gravity is so intense on a neutron star that a teaspoon of neutron-star stuff would weigh about a billion tons, a human would be crushed to the thickness of a postage stamp, and the entire population of Earth, if shipped to a neutron star, would be compressed into a volume about the size of an aspirin tablet.) Newly formed neutron stars must rotate extremely rapidly, with periods measured in milliseconds. And, finally, any magnetic field embedded in the original stellar core is amplified during the compression, reaching field strengths on the order of trillions of times the Earth's magnetic field (and even millions of times those in the hearts of solar flares). Strange objects, neutron stars represent states of matter unimaginably different from what we are used to.

After a supernova explosion has occurred, intense electric fields are established near the rapidly spinning, highly magnetised neutron star. These fields accelerate particles to extremely high energies, which in turn emit beams of often invisible (radio and x-ray) radiation that sweep past Earth each rotation period, thereby generating the flashes characteristic of "pulsars". Hundreds of pulsars were found during the past decade with radio telescopes, and the most famous one, the pulsar in the Crab Nebula supernova remnant several thousand light-years away, has been observed over the entire electromagnetic spectrum from radio waves to gamma-rays. Since both Asian and Middle-Eastern historical records clearly prove that this nebula originated in a supernova explosion in July, 1054 A.D., we can be virtually sure that at least this pulsar originated in a supernova explosion. In fact, we have little reason to doubt that all radio pulsars are neutron stars produced by supernova explosions, although, oddly enough, few supernova remnants have an observable pulsar within them. (Conversely, we are not surprised that most pulsars don't have supernova remnants associated with them, because pulsars probably endure for millions of years whereas supernova remnants disappear long before that.)

Potential Stellar Black Holes

A large fraction of all stars in the sky orbit another star as members of binary star systems. Observations of the behavior of the two members of a binary system permit us to infer some stellar properties that could not otherwise be determined. For example, precise timing of the radio pulses from a pulsar located in a binary system has provided an accurate value for the mass of the neutron star that causes the pulsar phenomenon.

Type I supernovae, characterized by optical emission that decreases slowly in the aftermath of the explosion, are probably descendants of white dwarf stars located in binary star systems of low mass. White dwarfs are carbon-rich, burned-out remnants of dead stars; the core of our Sun is destined to become such an object. They are produced when the core of a star somewhat more massive than the Sun succeeds in shedding its outer layers to reveal the core itself. The shedding occurs partly in the form of a stellar wind of a luminous cool star, and partly when a planetary nebula is slowly ejected into space. The white dwarf itself contains the ashes of the nuclear reactions that previously fueled the star during its lifetime. Provided the dwarf is alone, it is destined to become ever colder, eventually becoming a black dwarf – an invisible clinker in space.

According to one currently favored model, supernovae of Type I occur when matter flows from a companion star onto the white dwarf, increasing its mass above the limiting value of 1.4 solar masses; since burned-out stars having masses greater than this value must collapse, the white dwarf free-falls inward, thus triggering runaway nuclear reactions. This in turn causes an explosion that disrupts the white dwarf and produces large quantities of iron and iron-group elements. Observation of spectral features characteristic of iron in the spectrum of a supernova that appeared in another galaxy during 1972 lends support to this idea. Large amounts of the radioactive isotopes of nickel and cobalt are produced in the explosion and their subsequent decay over many weeks explains not only much of the light of type I supernovae but also the precise way in which it declines with the passage of time. Interestingly, a spectral feature attributed to radioactive cobalt has also been tentatively identified in the 1972 extragalactic supernova.

Like many other objects and regions in the Universe, binary systems also give rise to invisible events. One of the most important developments of the past decade was the discovery that compact, high-luminosity x-ray sources are located in binary star systems. The initial observations were made by the small satellite *Uhuru* (meaning "freedom" in Swahili), which was launched by NASA from a platform in the sea off the coast of Kenya. The x-ray emissions from these sources result from the intense heat generated as matter transfers from the atmosphere of a normal star to another object in orbit about it, and which is thought to be a collapsed star such as a white dwarf, neutron star, or black hole. The infalling matter does not fall directly into the collapsed star, as it is prevented from doing so by its tendency to rotate (angular momentum). Instead, the infalling matter forms a rapidly rotating disk of loose gas in orbit around the compact object. The matter in this so-called accretion disk slowly spirals in under the influence of friction, and becomes heated to temperatures up to tens of millions of degrees, before falling onto the collapsed object.

Perhaps the most famous binary x-ray source, Cygnus X-1, provides the best evidence for the existence of a stellar black hole – arguably the ultimate kind of invisibility in the Universe. Our Galaxy may well contain millions of stellar black holes, but we don't know for certain as we have no way to detect such isolated holes with current techniques. However, whenever stellar black holes are members of close binary systems, they should gravitationally pull matter from their neighboring companion, force the matter to collide violently with itself, and thus heat it greatly; to an external observer, these rather violent events should yield a telltale sign of intense x-ray emission as a sort of Last Hurrah before the stolen matter falls into the black hole, where it presumably remains trapped forever. This is precisely the case for Cygnus X-1, whose x-ray radiation implies that this stellar binary system (located some 6000 light-years from Earth) contains an unseen compact object of about 10 solar masses in orbit about an ordinary bright star companion. At this time we are driven to the conclusion that this unseen object is a black hole, as the theory of stellar evolution admits no other solution for an object of large mass which is also compact.

During the 1970s, satellite observatories demonstrated that many x-ray sources in binary systems pulsate rapidly, somewhat like the radio pulsars discussed earlier. These systems apparently contain rotating neutron stars having strong magnetic fields that channel matter accreting from a binary companion into the magnetic north and south poles of the star, heating it to x-ray temperatures as it slams into the stellar surface. The resulting x rays emerge in broad beams that sweep around like those of a lighthouse beacon as the neutron star rotates. Observations of such "x-ray pulsars" are providing valuable insight into the properties of the extraordinarily dense matter in neutron-star interiors.

In a surprising development, intense x-ray bursts were recently discovered from another major class of binary stars. Found near the central regions of our Galaxy and also near the centers of rich clusters of stars, these systems (called "x-ray bursters") emit thousands of times more radiation than our Sun, and in rapid bursts that last only several seconds. The bursts seem to arise from weakly magnetised neutron stars that are members of stellar binary systems of low mass. In these systems, matter accreted by the neutron star accumulates on or near the star to a depth sufficient to suddenly commence nuclear reactions (due to the pressure of overlying material). These reactions then release huge amounts of radiation in a burst of x rays; after several hours of renewed accumulation, a fresh layer of matter produces the next burst.

In an equally startling discovery, optical astronomers found that the object with the catalog name "SS 433" ejects more than an Earth-mass every year in two oppositely directed narrow jets of gas moving at approximately one fourth the speed of light. Periodic changes in the optical emission spectrum imply that the jets precess like a spinning top, tracing out a complete cone twice a year. This interpretation has been recently confirmed by high-resolution radio images of SS 433 obtained by means of very-long-baseline interferometry, which demonstrate the invisible helical patterns of matter sprayed into interstellar space as the jets twirl around like a fireworks display. The basic mechanism of SS 433 is also thought to be a binary system containing a compact object; however, the origin of this peculiar system, the mechanisms that accelerate and align the matter into jets, and the cause of the precession are still unclear.

In an even more dramatic discovery, several military satellites detected gamma-ray bursts in space – bright, irregular flashes of gamma rays lasting only a few seconds. Although the poor spatial resolution of the gamma-ray detectors has precluded the optical identification of most of the gamma-ray sources, they probably resemble scaled-up versions of binary-star systems in which matter, falling onto a neutron star, experiences a sudden flash of thermonuclear burning. On the other hand, the process may not really be that simple, for, on March 5, 1979, the most intense gamma-ray burst ever recorded was observed from the direction of a known supernova remnant in the Large Magellanic Cloud; provided it really did originate at the 150,000 light-year distance of this region beyond the outskirts of our Milky-Way, the intrinsic luminosity of the burst would be so great as to defy explanation by any mechanism now known to astrophysics.

Active Galaxies and Quasars

A few decades ago, the pioneers of radio astronomy discovered a source of unusual activity in the constellation Sagittarius. That source, called Sgr A, is now known to be located at the exact center of our Milky Way Galaxy. Optical studies of the galactic center region are prevented by obscuration due to interstellar dust, but in the late 1970s observations of radio and infrared radiation (which can penetrate the dust) revealed abnormal emission features arising from ionized gas in the vicinity of Sgr A. The abnormality arises from the fact that the spectral features extend over an especially broad wavelength range. If, as seems likely, the broadening of these features results from a rapid rotation of the emitting gas (so that high approach and recession velocities tend to broaden the features by means of the Doppler effect), then our galactic center is the site of a vast whirlpool or vortex containing thousands of solar masses spread over several light-years. To prevent this rapidly spinning vortex from dispersing outward, a strong gravitational field is needed to keep the whirlpooling matter in orbit. The simplest solution is that this gravitational field is caused by several million solar masses concentrated in a very small region at the center of the whirlpool. Moreover, interferometric radio observations using intercontinental baselines have shown that Sgr A is smaller than our Solar System; if this is the mass concentration itself, it could simply be an unusually dense cluster of stars. On the other hand, because Sgr A is such a violent singular object containing relativistic electrons, we might more reasonably suppose that we are dealing with an accretion disk that produces the relativistic electrons responsible for the radio emission of Sgr A. If so, the object at the center of the disk could well be a supermassive black hole. Such an object would be only a distant cousin of the stellar black holes we have described earlier. The latter form in the normal course of evolution of individual stars, but supermassive black holes must have formed by the collapse of a giant gas cloud, or perhaps by the collisional merging of millions of stars.

In even more recent observations of our galactic center, equipment on board both balloons and satellites has detected a strong emission feature in the gamma-ray part of the spectrum. We suspect that this feature arises from the annihilation of matter and antimatter, specifically between electrons and their antimatter-opposite positrons; such annihilations release the full energy equivalent in the masses of the two particles. Further observations show that the gamma rays decreased in strength by more than a factor of two within a six-month period, implying that the source of the positrons must be extremely compact; otherwise, the delays experienced by radiation coming from more distant parts of the source would prevent the weakening observed. We would not be surprised if this source turns out to be associated in some way with the suspected supermassive black hole at the heart of our Galaxy.

The center of our Galaxy is truly the site of unexpected violence, despite the relative quiescence that we enjoy in the galactic suburbs, some 30,000 light-years away. Whirlpools of hot gas, antimatter, and supermassive black holes were supposed to be part and parcel of other, more exotic galaxies far away – or so we thought until recently.

There is no doubt that some distant galaxies are extraordinarily active, often displaying phenomena far more violent than anything in our Milky Way. However, because of their much greater distances, the remote galaxies are more difficult to resolve spatially. For example, for some time we have known of a bright "spike" of emission at the center of the neighboring Andromeda Galaxy – a region that probably harbors an abnormally dense concentration of stars. Some peculiar spiral systems, called Seyfert galaxies, also exhibit violent activity in their centers, including radio and x-ray emission, rapid motion of ionized gas, and powerful infrared emission concentrated within their innermost few hundred light-years; just as we argued regarding the gamma-ray source at the center of our Galaxy, the variability of the Seyferts' x-ray emission indicates that most of their intense radiation must arise in regions less than about a light-year across. Some elliptical systems, called radio galaxies, exhibit similar phenomena, and in addition display vast lobes of radio-emitting material extending about a million light-years or more into intergalactic space. These lobes appear to be fed by jets of relativistic particles emerging from the active centers we observe in the radio galaxies. For example, the giant radio galaxy M87 in the Virgo cluster has a bright and compact optical, radio, and x-ray source at its center, from which a jet of matter emerges. Analyses of the stellar distribution in this region suggests that this galaxy houses a concentration of about a billion solar masses within its central few hundred light-years. Such a concentration cannot be explained by an abnormally dense cluster of stars, so some new phenomenon is needed. Is it a supermassive black hole? And, in another famous example, the center of the radio galaxy Centaurus A emits a large fraction of its power as gamma rays, a phenomenon entirely uncharacteristic of stars, which emit mostly in the visible part of the spectrum.

As if the active galaxies were not troublesome enough, the objects called quasars are even more difficult to explain. Quasars are starlike objects whose spectral features have large Doppler shifts indicative of large recessional velocities, and hence, large distances in the expanding Universe.* The largest Doppler shift thus far observed for a quasar corresponds to a velocity of 91 percent of the speed of light. Although the first quasars were discovered by analyzing the optical spectra of visible objects found at the same positions as pointlike radio sources, many new quasars were recently found by examining the optical counterparts of faint x-ray sources discovered by the *Einstein* orbiting x-ray observatory. While their radio, infrared, and x-ray, as well optical emission, resemble those of the centers of Seyfert galaxies, quasars pose the problem in its most acute form because they emit so much more powerfully.

For nearly two decades, some astronomers have doubted the origin of the quasars' large Doppler shifts. However, this long-standing controversy is now close to resolution, for recent observations demonstrate that many quasars reside among groups of galaxies having essentially the same Doppler shift; since we know that these shifts are valid indicators of the distances to normal galaxies, these findings lend support to the idea that quasars really are at the vast distances implied by their Doppler shifts. At these distances – typically more than several billion light-years away – the energy released by a quasar is equivalent to the complete conversion into energy of an amount of matter equal to that in our Sun *every year.* That makes quasars, aside from the big bang itself, the most powerful explosions in nature.

* As we shall discuss in Chapter 7, nearly every object in the Universe is moving away from us; in fact, the velocity of recession is proportional to the distance of the object.

Quasars resemble radio galaxies in that compact, active regions are connected by jets to distant radio-emitting lobes of gas. As best we can tell, the basic energy source in quasars must reside in the central compact region that most astronomers assume may well be the core of a galaxy. High-resolution radio observations using the technique of very-long-baseline interferometry have revealed small, jetlike structures emanating outward from the active regions of many quasars; these small jets of matter, perhaps a few light-years in length, are aligned with the much larger jets of matter that often extend up to millions of light-years, thus establishing a causal connection between the outwardly bound jets of relativistic particles and the quasar cores where they are almost certainly generated. Indeed, virtually all the observations made of the compact radio sources imply that electrons are impulsively injected into and then magnetically trapped in the huge outlying lobes. "Impulsive" is not too strong a word here, for in many cases blobs within these small radio-emitting jets appear to travel outward with speeds close to that of light itself. Incredibly, for a few quasars the blobs seem to move faster than the velocity of light (which, of course, is strictly forbidden in the currently known fabric of physics); thus far, theorists have been able to squirm out of this dilemma without requiring faster-than-light velocities, provided the large inferred velocities are an illusion produced by blobs moving nearly toward us at slightly less than light speed.

Many new quasars were recently discovered solely by means of their x-ray emission. The nature of their spectra implies that the x rays are emitted either by extremely hot (ten-million-degree) plasmas, or by relativistic particles spiralling within magnetised regions. Some x-ray quasars vary on time scales as short as a few hours, implying that their source regions are smaller than our Solar System – despite the fact that the power radiated in some cases exceeds a trillion solar luminosities. We now realize that the distant, x-ray emitting quasars are so numerous that they must account for a substantial fraction (and possibly all) of the previously unresolved x-ray background that floods the entire sky. At least one quasar, 3C 273, is known to be a strong gamma-ray source as well as a potent emitter of radiation in virtually all other parts of the electromagnetic spectrum.

Theoretical models of quasars postulate a layered structure surrounding a compact central energy source. Farthest out are the clouds of cool gas that were ejected by the quasar at earlier epochs and that are now detected through observations of their heavy element-rich absorption features. Inside, emission features are formed; these features, somewhat like those arising in the central regions of normal galaxies, including our own Milky Way, extend over a broad wavelength range, suggesting that numerous clouds revolve in rather small orbits around a central region housing a very large mass. At the center is a compact energy source responsible for the acceleration of particles and for the generation of gamma rays and x rays, as well as optical and perhaps infrared radiation, by a complex combination of processes that we have yet to fathom in detail.

Because of the vast amounts of energy produced in such small volumes of space, the nature of the central engine within quasars is naturally of great interest. Among leading candidates are a compact cluster of neutron stars or stellar black holes – or both, intermixed – undergoing frequent collisions, a massive cloud of plasma stabilized by rotation and inundated with magnetism, and an accretion disk formed from matter spiraling in toward a single supermassive black hole housing perhaps billions of solar masses. Though a consensus has not yet been reached within the astronomical community, many researchers are leaning toward the last of these possibilities.

Do quasars and active galaxies really harbor supermassive black holes in their hearts? If so, how did they form, and what is the source of matter that feeds the black holes? How is the gravitational energy of the accreting matter converted into the observed radiation? These are among the most exciting unsolved problems in astronomy today.

Future Prospects

Current theories of violent events in the cosmos, ranging from solar flares to quasars, offer a variety of possible explanations. In this section, we suggest how several new observational facilities can help us choose among the many possibilities and thus to advance our knowledge of these demonstrably unearthly phenomena.

The mechanisms underlying the acceleration of particles in many astronomical contexts can be studied up close by observing the flaring activity of our Sun. Though not as violent as black holes, galaxy centers, or quasars, the basic physical process occurring in solar flares is thought to mimic quite closely those of the more active regions. Accordingly, toward the end of the decade, the Sun is scheduled to be investigated with a versatile assembly of instruments, especially those aboard the *Advanced Solar Observatory* (*cf.*, Appendix E). Current design stipulates the heart of this satellite to be a 1.3-meter diameter mirror capable of (optically) resolving features as small as 0.1 arc second, or 70 kilometers on the Sun. Ultraviolet and x-ray telescopes aboard this orbiting observatory will also yield high-resolution measurements of temperatures and densities in the flaring material, and gamma-ray detectors will provide information on the properties of the highest-energy particles accelerated by the flares. Thus, as we enter the 1990s, we should have in hand much new information about solar plasma, much of it obtained with unprecedented resolution before and after solar flares.

Questions regarding the sources, acceleration, and propagation of cosmic rays in our Galaxy will also receive added impetus in the decade ahead. In particular, with the deployment of highly sensitive detectors on board satellites operating in the gamma-ray region, we shall be able to make improved measurements of the relative numbers of different atomic nuclei in the cosmic rays, which should in turn provide insight about their origins. The detection of gamma rays is relevant here because this kind of radiation is released when cosmic rays interact with interstellar gas; since the gamma rays have no charge, they are unaffected by interstellar magnetic fields, and thus should indirectly yield information about the (charged) cosmic rays which the cosmic rays themselves cannot. Of special import, the *Gamma-Ray Observatory* (*cf.*, Appendix I) will monitor gamma-ray emission from supernova remnants, and thus provide crucial tests of theories for the origin of cosmic rays in supernova explosions as well as their acceleration by shock waves.

Our best ideas about the formation of heavy elements will also be tested by the *Gamma-Ray Observatory*. Its capabilities should permit detection of gamma-ray spectral features emitted by the nuclei of chemical elements freshly synthesized in supernova explosions as far away as the Virgo cluster of galaxies, where a new supernova occurs about once a year. The dynamics of such explosions will be examined by observing the changes in their x-ray and ultraviolet spectra with the *Advanced X-Ray Astrophysics Facility* (Appendix D) and the *Space Telescope* (Appendix A), respectively.

The phenomena displayed by compact stellar objects are so rich and varied that their study will require the concerted use of many of the newly proposed instruments. For example, the *Very Long Baseline Array* (*cf.*, Appendix H) will be able to observe the dynamics of jets emerging from galactic objects such as SS 433 with unprecedented precision; the array's many different baselines will ensure observations with an order-of-magnitude greater contrast than heretofore possible. The mechanisms for particle acceleration and radiation in pulsars will become clearer from observations of their x-ray and gamma-ray emissions with the *Advanced X-Ray Astrophysics Facility* and the *Gamma-Ray Observatory;* by extending our studies to fainter objects, these satellites will, among other objectives, advance our understanding of accretion onto compact objects, the physics of matter and radiation in superstrong magnetic fields, and the interior structure of neutron stars. In fact, because of its much greater ability to detect faint sources of x rays than previous instruments, the *Advanced X-Ray Astrophysics Facility* will permit us to observe binary x-ray sources in nearby galaxies, thus providing clues to their origin and evolution.

The phenomena of transient x-ray and gamma-ray sources will also be observed with unprecedented sensitivity by the *Gamma-Ray Observatory*. A particularly outstanding challenge is to detect and identify the optical and x-ray counterparts of the mysterious gamma-ray burst sources; this will require a coordinated program of observations involving gamma-ray and x-ray telescopes in space and optical telescopes on the ground.

As for quasars, many of our new instruments will have powerful capabilities for advancing knowledge. For example, because its high spatial resolution should permit us to detect the parent galaxies of quasars unambiguously, the *Space Telescope* should help us determine if the quasars are really located within the centers of galaxies as most astronomers suppose. And if so, this orbiting observatory should also permit us to determine whether the quasar-associated galaxy is a spiral, an elliptical, or another type, thereby furnishing key information for theoretical models of quasars. Moreover, the increased sensitivity of the *Space Telescope* will allow searches for groups or clusters of galaxies neighboring quasars out to much greater distances than can any ground-based optical telescope. However, equipment with the power of the *New Technology Telescope* (*cf.*, Appendix F) will be needed to measure their Doppler shifts and thus estimate their distances.

Our understanding of the regions causing the emission and absorption features in the spectra of quasars will advance greatly as a result of long-duration observations in space. For example, the ultraviolet instruments to be carried aloft by the *Space Telescope* are designed to detect even small amounts of gas in the outlying regions (halos) of galaxies or in intergalactic clouds situated along the line of sight between us and the quasars. The resulting knowledge of the physical conditions in the intervening regions should determine whether the absorption occurs in blobs ejected by the quasars or in unrelated clouds; if in the latter case the clouds should prove to be truly intergalactic in nature, then the ultraviolet observations should also yield the relative abundance of hydrogen and helium in gases that have not yet been contaminated by helium synthesized in stars – a key datum for cosmology.

Of greatest interest, perhaps, the mystery of the central energy source that powers quasars and active galaxies will be addressed in new ways by observations with the *Gamma-Ray Observatory*, the *Shuttle Infrared Telescope Facility*, and the *Submillimeter-Wave Radio Telescope*. For the first time, instruments sensitive to short-wavelength gamma rays all the way down to long-wavelength radio waves will be able to sample simultaneously the energy emitted by quasars and active galaxies. Such multiple observations are particularly important because much of their total emission may emerge in the gamma-ray, infrared, and radio bands of the electromagnetic spectrum; and since the emission is often variable, simultaneous observations are needed. Furthermore, if, as seems likely, the highest energy emissions occur progressively closer to the central energy source, observations of rapid time variations of x-ray emission with the *Advanced X-Ray Astrophysics Facility* will likely provide vital clues to the nature of these sources.

Additional clues of a qualitatively different nature will be secured by observations at radio wavelengths. The *Very Long Baseline Array* will provide the highest spatial resolution at any wavelength, amounting to, for relatively nearby quasars, a mere 1 light-year. Such observations will enable us to distinguish, for example, between the precisely constant location mandated if the central engine is singular in nature (a black hole) from the variably flaring localities predicted by a model based upon recurrent supernovae. Thus, we face the exciting prospect of being able to directly image the central sources, and perhaps the prime engine itself. Such observations should also provide direct evidence as to how particles are accelerated to high energies and ejected in beams that extend far into intergalactic space.

Concluding, we now have much evidence suggesting that the Universe harbors a wide but interrelated continuum of activity, of which the quasars represent but the most extreme example. Despite the vast energies involved in some of the outburts characterizing many of the most active objects, and despite few firm conclusions at hand as to how the details of the mechanisms work, a certain confidence has recently emerged that we can now devise dynamically self-consistent and *testable* models of cosmic violence. Accordingly, we can currently cite no compelling reason to suspect that any "new physics" is needed to understand the many faceted objects displaying activity.

Through our invisible images, we are gradually recognizing that much of the Universe, from our perspective, is normally violent. Just as Galileo argued that Nature is less immutable than Aristotle once thought, we are only beginning to realize that Nature is apparently more active than even Galileo once thought – or observed.

The Supernova Alert Team

I am fascinated by the notion that, almost certainly, a viewable star has already exploded as a supernova, but the radiation from this stupendous event has yet to reach our planet. Should such a supernova suddenly appear in the sky, we can be sure that all the world's major astronomical instruments will immediately focus in the direction of this, the grandest of light shows.

I recall about ten years ago, when I was a staff member of the Harvard-Smithsonian Center for Astrophysics, a celestial object suddenly brightened by a large factor, and some of us thought sure that we would soon have the opportunity to study a nearby galactic supernova for the first time since the invention of the telescope nearly four centuries ago. George Field, then director of the Center, had earlier established a Supernova Alert Team comprised of several observationalists who were prepared to commandeer every available ground-based telescope and orbiting spacecraft at our disposal. We were charged with the study of the supernova's radiation at a variety of different wavelengths across the entire electromagnetic spectrum, and, of utmost importance, we would be eager to establish if the newly released radiation might be harmful to life on Earth.

As fate would have it, the brightened object, known as Cygnus X-3, began flaring at the start of a long Labor-Day weekend, so our team was well scattered on vacation; after all, we hadn't expected a candidate supernova to inconsiderately time its earthly appearence on a human holiday! As I had just ended a long observing run on MIT's Haystack radio telescope, I was, though exhausted, well positioned to go back on the telescope. Along with several colleague scientists, I began monitoring the object's violent activity. But we soon encountered a problem that plagues radio astronomy worldwide, and increasingly so with the incessant advance of technology. Our broad-based measurements of Cyg X-3 were harmed by interference from all sorts of electromagnetic signals generated by our high-tech society, especially those produced by military communications and radar monitorings. Today the airways are flooded with electromagnetic garbage, and it's only a matter of another couple of decades before most ground-based radio astronomy is forced to halt operations as strong man-made signals overwhelm the weak radio radiation naturally emitted by cosmic objects. This interference is the cause of much of the static on your home AM radio receiver.

At any rate, if Cyg X-3 was to be the first nearby supernova since Renaissance times, we were desperate to make an accurate measurement of its rising activity. So my colleagues and I took the bold step of requesting help of some senior officials at MIT's nearby Lincoln Laboratory. Our "plea for quiet" rapidly filtered up the chain of command to some unnamed person or persons in Washington. And somewhat surprisingly, we were soon told that, at a specific time and for a period of exactly eight minutes, we would have an opportunity to measure Cyg X-3's radiation virtually free of interference. I was a bit skeptical that we would see any drastic improvement in quality of reception, but, as the scheduled time neared I nonetheless prepared our telescopic equipment to gather as much cosmic data as possible. And was I wrong, for at exactly the appointed time, nearly all the electromagnetic interference disappeared; even the underlying static that had plagued every radio observation I had ever made until that moment (or since) became less. The Earth, it seemed, had become as radio-quiet as it had naturally been in the days before Marconi. It was a radio astronomer's paradise!

I often wonder what might have happened if the Soviets had initiated a first strike during those eight minutes, when, apparently, much of the American military radar systems had been turned off (or at least greatly decreased in transmission power). Not to worry; at precisely the end of the agreed time interval, all the interference returned and the static resumed its normal hiss. Even so, during those eight minutes, thanks to exemplary cooperation between science and government, the Cyg X-3 object had been measured with considerable accuracy.

As things turned out, our observations and those of legions of other groups worldwide proved that this object was not a supernova, but merely a somewhat unstable stellar system experiencing a sporadic outburst of energy. A decade ago, this kind of event was new and unexpected; now, as discussed elsewhere in this book, we recognize that cosmic activity seems more the rule than the exception. Thus, we still await radiative news of a grand explosion – a genuine supernova in our galactic neighborhood – and one that will once again require the close cooperation of science and government for the good of all human life on Earth.

E.J.C.

7
Universe
Its Large-Scale Structure

Though the difference between men and other animals is enormous, yet one might say reasonably that it is little less than the difference among men themselves. What is the ratio of one to a thousand? Yet it is proverbial that one man is worth a thousand when a thousand are of less value than a single one. Such differences depend upon diverse mental abilities, and I reduce them to the difference between being or not being a philosopher; for philosophy, as the proper nutriment of those who can feed upon it, does in fact distinguish that single man from the common herd in a greater or less degree of merit according as his diet varies.

He who looks the higher is the more highly distinguished, and turning over the great book of nature (which is the proper object of philosophy) is the way to elevate one's gaze. And though whatever we read in that book is the creation of the omnipotent Craftsman, and is accordingly excellently proportioned, nevertheless that part is most suitable and most worthy which makes His work and His craftsmanship most evident to our view. The constitution of the universe I believe may be set in first place among all natural things that can be known, for coming before all others in grandeur by reason of its universal content, it must also stand above them all in nobility as their rule and standard.

From the Dedication to the *Dialogue Concerning the Chief World Systems,* by Galileo Galilei, Florence, 1632

Like Motes in a Sunbeam, galaxies drift mysteriously in the depths of space. Everywhere they are clumped into groups and clusters millions of light-years across, with an occasional supercluster of galaxies spanning up to hundreds of millions of light-years. However, on even larger distance scales, the groups and clusters seem to be distributed at random in space, the number of galaxies in a given large volume of space being about the same throughout the Universe. Succinctly stated, on scales of order several hundred million light-years and greater, the Universe appears to be quite smooth and structureless. Studies of matter and radiation on the largest scales of all comprise the subject of cosmology.

Probing the Largest Scales

The uniform spread of matter on very large scales invites comparison between the observations and a simple theoretical model of the Universe (or cosmological model) derived in 1922 from Einstein's general theory of relativity. According to this model, the geometry of space itself is curved by matter, and this curvature forces matter to move; at any epoch the Universe must be either expanding or contracting. In 1929, the American astronomer Edwin Hubble demonstrated that the system of galaxies is actually expanding as expected from the model. He did this by measuring the velocities of galaxies from the Doppler shift of the lines in their spectrum; he found that not only are the shifts almost always toward the red (or longer wavelengths), indicating velocity of recession, but also that the velocity is proportional to distance, a relationship also predicted by the model. The ratio between velocity and distance is called the Hubble constant; its value is uncertain, but is estimated to be between 15 and 30 kilometers per second for each million light-years of distance.

The fact that the Universe is actually expanding as predicted forces us to confront a bizarre implication of the cosmological model: the Universe must have originated some 10 to 20 billion years ago in a powerful explosion – termed the big bang – before which neither time nor space had any meaning.

Like archeologists who dig through ancient rubble in search of hints about the origin and evolution of culture, today's astrophysicists probe the history of the Universe by studying the radiation from objects so distant that it left them in the remote past, when the Universe was very different. With the time machines we call telescopes, we have found objects so distant that they recede from us at more than 90 percent of the speed of light – apparently the debris of a huge explosion that once was. Radio antennas have uncovered faint signals widely considered to be remnants of the big bang –

an omnipresent radio static (termed the cosmic microwave background) that pervades all space, bathing us in the cooled relic of creation itself. And nuclear reactions occuring during the first few minutes of existence produced key chemical elements (such as hydrogen, helium and their isotopes) whose abundances have been measured in space by various ground-based and orbiting instruments. Amazingly, all the data agree with the theoretical predictions based upon a particular big-bang model.

For these reasons, the big-bang model has become our standard cosmological idea with which to compare observations. This is not to say that the model is completely correct, for the data are imprecise, their interpretation might be in error, and the theory could be wrong. A central task for the future is the further development of the big-bang model as well as its continued testing against all available observations.

As noted above, one of the foremost stipulations of the big-bang model is that matter be distributed uniformly on the largest scales. By using a variety of approaches, astronomers are now subjecting this prediction to more stringent tests. One approach is to locate galaxies in three dimensions by deriving their distances from the Doppler shifts in their spectra. Because in the past much time has been needed to record the spectrum of a galaxy photographically, not many distances to galaxies have been determined this way. Now, however, electronic detectors have greatly speeded up the process, even on telescopes of moderate size. Using this data, astronomers have confirmed the existence of superclusters of galaxies and of large voids apparently devoid of any galaxies between them. With a telescope whose aperture is 4 to 5 meters, we should be able to measure the distances of many galaxies out to a billion light-years, and to test the notion that the galaxies are indeed spread uniformly on that scale.

X-ray and gamma-ray astronomy also tell us something about the large-scale distribution of matter. A diffuse background emission not attributable to any known sources appears in each of these parts of the electromagnetic spectrum; since we receive the emission uniformly from nearly every direction (that is to say, "isotropically"), it cannot originate within our Galaxy, but must instead arise at distances comparable with the size of the Universe itself. Toward the end of the 1970s, an orbiting spacecraft established that the x-ray background agrees with the radiation expected from a gas having a temperature of nearly a half-billion degrees Celsius, suggesting that such superheated gas may be spread throughout the Universe. More recently, on the other hand, the *Einstein* x-ray observatory discovered that individual quasars at large distances are powerful x-ray sources – so powerful, in fact, that quasars at even larger distances than can be currently detected individually must account for a substantial fraction of the observed x-ray background. As some quasars have also been found to be powerful gamma-ray sources, the gamma-ray background might also be the accumulation of huge numbers of quasars. Thus, the x-ray and/or gamma-ray backgrounds arise (at least partly) from very large numbers of very distant quasars, too faint to detect individually. We are still unsure, however, how the spectra of these quasars would sum up so as to mimic the spectrum of a hot gas. Whatever the detailed explanation, the x-and gamma-ray backgrounds are highly isotropic, which seems to require a uniform distribution of whatever causes them. This would be in accord with the big-bang model.

The cosmic microwave background radiation also provides information about the large-scale structure of the Universe. For instance, measurements of its intensity over the sky have revealed a smooth variation that arises from Earth's motion through the cosmos. The observed variation is unexpectedly large, corresponding to a velocity of some 500 kilometers per second for the group of galaxies of which our Milky Way is a member. These same measurements reveal that any other variations must be very small, implying that the early Universe must have been highly uniform. Ground-based measurements of the radiation at wavelengths longer than 1 millimeter indicate that the spectrum of the radio background radiation does not deviate significantly from that of thermal emission, as predicted by the big-bang model. The *Cosmic Background Explorer* satellite to be launched by NASA in the late 1980s will make definitive studies of both the spectral and the directional distribution of the microwave background radiation.

Expansion Time Scale

As noted earlier in this chapter, the big-bang model predicts that galaxies should recede from one another with velocities proportional to their separations; the factor of proportionality is the Hubble constant. The big-bang model also predicts that the reciprocal of the Hubble constant (often termed the "Hubble time") roughly equals the present age of the Universe – that is, the time elapsed since the big bang.

To determine the value of the Hubble time, we need to measure the distances of remote galaxies, using a "ladder" of interconnected distance scales established by different methods; each step of the ladder enables us to probe further into space. When Hubble first studied the recession of the galaxies in the 1930s, the data implied that the Universe was some 2 billion years old. This Hubble time has since been revised several times – to 5, then to 10, and then to 20 billion years. Each revision resulted from a major advance in our understanding of the properties of stars or galaxies that are used to construct the ladder of cosmic distances.

The value of the Hubble constant is one of the most important numbers in contemporary astrophysics; the Hubble time enters all cosmological calculations in a fundamental way. To find its true value, we need to be sure of each step of the distance ladder; in particular, any contributions to the velocities of the galaxies that are not due to the expansion of the Universe must be carefully considered. For example, when the motion of our local group of galaxies was revealed by studying the cosmic background radiation (as noted in the previous section), a more consistent set of data for the Hubble time emerged. We now feel confident that the age of the Universe lies between 10 and 20 billion years. Although the factor of two uncertainty is not too bad, considering the difficulty of making the measurements on very distant objects, it is crucial to improve our accuracy.

Hidden Mass and the Fate of the Universe

For nearly two decades now, astronomers have been increasingly puzzled by what might be called the "hidden mass" problem, according to which most of the matter comprising the Universe is apparently invisible. Spectra of individual galaxies indicate that, like our own Milky Way, they contain normal stars; however, the motions of various stars within galaxies are so large that galaxies would soon disperse into intergalactic space if the only gravitational attraction holding them together were that of the stars we see. Additional mass must be present in some form that is hidden from our immediate view – enough to supply the gravitational force needed for stability. Similar results emerge from studies of groups and clusters of two or more galaxies: their masses must be at least ten times greater than the masses of all the visible stars in them.

What is the nature of the matter comprising the hidden mass? In principle, it could be made of diffuse gas, collapsed stars such as white dwarfs, neutron stars, and black holes, or faint red dwarfs. Some researchers have even suggested some unknown form of matter such as huge numbers of neutrinos that have finite rest masses. To be sure, theorists have shown that the conditions were ripe for the production of large amounts of neutrinos in the aftermath of the big bang. If their rest mass is zero, they would not contribute significantly to the hidden mass. Conversely, recent advances in theoretical particle physics indicate that their rest mass might indeed be finite. If so, neutrinos could account for at least the hidden mass in clusters of galaxies, if not also for the hidden mass within galaxies themselves. An important task for experimental physics is therefore to determine whether or not neutrinos do possess any finite rest mass; at present the evidence is not compelling either way.

Recent radio, optical, and x-ray observations have shown that hundred-million-degree gas exists within clusters of galaxies, but the amounts of such gas are insufficient to account for the hidden mass in such clusters. Collapsed stars of various types could in principle comprise much of the hidden mass, but, according to the theory of stellar evolution, such stars are the descendants of the most massive stars and so would dominate the total mass only if, at early epochs, massive stars were the most frequently formed stars. Just the contrary is observed in our Galaxy, at least for those stars that have formed near the Sun; faint red dwarfs, which have among the lowest mass of any stars, are so numerous that they account for most of the mass bound up in stars. Perhaps faint red dwarfs themselves account for the hidden mass; large numbers of them in the outer parts (or halos) of galaxies would be consistent with the lower light levels observed there.

The hidden-mass problem is of crucial importance to modern astrophysics not only because it renders incomplete our inventory of matter in the Universe but also because it is intimately connected with the question of the ultimate fate of the Universe. According to our big-bang model, the Universe will continue to expand forever if the amount of matter in it is less than a certain critical value; from the observed rate of universal expansion – the Hubble constant – we calculate the critical density to be approximately 10^{-29} gram in each cubic centimeter. If the amount of matter exceeds this critical value, the current expansion will reverse at some time in the distant future, and the Universe will collapse back into a singular state much like that from which it began in the big bang. As we explained earlier, the abundances of some key elements and isotopes produced in the immediate aftermath of the bang (such as helium and deuterium, a heavy isotope of hydrogen) depend sensitively on the amount of ordinary matter present in the first few minutes. Observations of the abundances of these elements in interstellar space of our Galaxy, taken at face value, suggest that the mean density of the cosmos is only ten percent of the critical amount needed to halt universal expansion. However, this test is sensitive only to ordinary matter, and does not include any massive neutrinos since they would not have participated in the nucleosynthesis of helium and deuterium in the early Universe. On the other hand, current observational estimates of the matter – of all types, including massive neutrinos, perhaps – aggregated within groups and clusters of galaxies are as high as forty percent of the critical value. Since this value is larger than the one obtained from observations of helium and deuterium, we conclude that much of the matter gravitationally binding galaxy clusters *might* be in the form of elusive neutrinos rather than ordinary matter. If so, invisible subatomic neutrinos could conceivably comprise most of the matter in the Universe. We shall return to this intriguing possibility in the final chapter.

Future Prospects

In addition to the virtual certainty that Doppler-shift surveys of much more distant galaxies will give us a better idea of their spatial distribution in the decade ahead, we also anticipate that new instruments will improve our studies of the x-ray and gamma-ray background emissions that invisibly flood the Universe. For example, the *Advanced X-Ray Astrophysics Facility* (*cf.*, Appendix D) will be able to observe sources some hundred times fainter than can our currently best x-ray equipment; we should thus be able to determine whether faint quasars account for the well-observed but still mysterious x-ray background. Likewise, measurements of faint quasars by the *Gamma-Ray Observatory* (*cf.*, Appendix I) should provide similar information on the gamma-ray background. If we can prove that the x-ray and/or gamma-ray backgrounds are actually due to quasars, the fact that the background is highly isotropic means that the matter at great distances is distributed very uniformly. If, on the other hand, we are forced to conclude that intergalactic gas is responsible for at least part of the x-ray background, then we shall be able to infer that this gas is spread uniformly throughout space; moreover, the amount of gas inferred from the data will be an important datum for the theory of the evolution of galaxies, as discussed in Chapter 5.

Refinement of the cosmic distance ladder will require much more work. Development of high-tech methods for measuring stellar positions should permit a more accurate measurement of the distance to the Hyades star cluster in our Galaxy, one of the crucial first steps in the ladder of cosmic distances. Because of its ability to detect extremely faint objects, the *Space Telescope* (*cf.*, Appendix A) will for the first time be able to resolve variable stars in the Virgo Cluster, thereby eliminating an uncertain intermediate step of the distance ladder. The continued deployment of advanced optical detectors at ground-based telescopes will make possible the rapid measurement of Doppler shifts of galaxies, extending beyond the peculiar motions of galaxies in our own tiny corner of the Universe, and thus reaching out to regions where the velocities of galaxies are almost entirely due to the expansion of the Universe; the *Space Telescope* will be able to determine the distances of the same galaxies by comparing the brightness of their star clusters with the brightness of those in the galaxies of the Virgo Cluster, whose distances are more accurately known.

Solution of the hidden-mass puzzle is a major goal of astronomy in the decade ahead. Our first task is to map the distribution of the invisible matter in space. One way to do this involves studying the velocities of the star clusters in the neighborhoods of galaxies; since the motions of such clusters reflect the strength of the local gravitational field, we should be able to infer the spread of mass in their parent galaxies. To undertake such spectral measurements of stellar velocities of faint stars in even some of the nearest galaxies, we shall need a very large telescope such as the currently envisioned *New Technology Telescope* (*cf.*, Appendix F). As for the source of the hidden matter, our new equipment should be able to address directly several of the possible candidates. For instance, we might well be able to test if the faint red dwarf stars are the principal source of hidden matter by using the recent discovery that such objects are surprisingly intense sources of x rays, owing to the hot gases they retain in their envelopes, much as the Sun displays a corona. The *Advanced X-Ray Astrophysics Facility* should be able to detect red dwarfs by observing their accumulated coronal x-ray emissions, provided there are large enough numbers of such dwarfs.

On larger distance scales, the galaxies themselves can serve as probes of the distribution of matter within clusters and superclusters of galaxies. Since galaxies are much brighter than star clusters, research on galaxy clustering is already underway with telescopes of intermediate size. Even so, measurements of velocities of galaxies in more distant clusters are essential if we are to determine how the distribution of hidden mass has changed over the course of time; such studies will require observations with equipment much more sophisticated than we now have, in fact apparatus comparable to the planned *New Technology Telescope.*

Finally, our efforts to determine the ultimate destiny of the Universe will be enhanced in a variety of novel ways. One such way will probe in much greater detail the abundances of helium and deuterium synthesized in the early Universe. Recall, as noted above, that the currently observed amounts of these species suggest that the total quantity of matter is too low by a factor of ten for its gravitation to be able to halt the expansion of the Universe. However, the issue is clouded, for we know that some helium has been produced and some deuterium has been destroyed in stars during the roughly 10-billion-year lifetime of our Galaxy, so present abundances of helium and deterium in the Galaxy may not be closely mimicking their relative abundances in the primordial gas that emerged from the big bang. Our new tests will center about the various gas abundances in intergalactic space, if it exists, which should be primordial. As noted in the previous chapter, the spectra of distant quasars display absorption lines that probably originate either in clouds formed by the outward ejection of thick shells of gas from

Continued on page 135

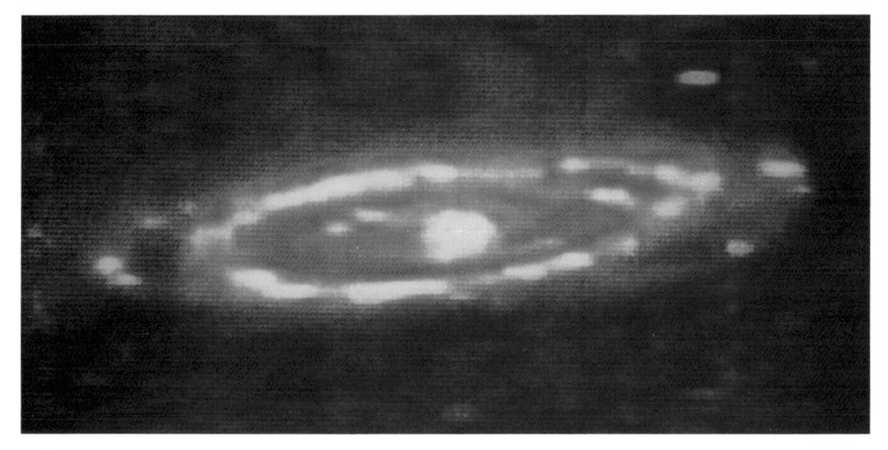

(above) M31, the Andromeda Galaxy, at infrared wavelengths as observed by the recently launched Infrared Astronomical Satellite. This image was taken by having the satellite repeatedly scan the galaxy to improve both the resulting spatial resolution and sensitivity. The enhanced color image shows radiation at 12 microns in blue, 60 microns in green, and 100 microns in red. The satellite galaxy NGC 205 shows as a small smudge directly above the center of main galaxy. A much hotter and hence bluer galaxy appears below the left edge of Andromeda.

(below) Caption on page 128

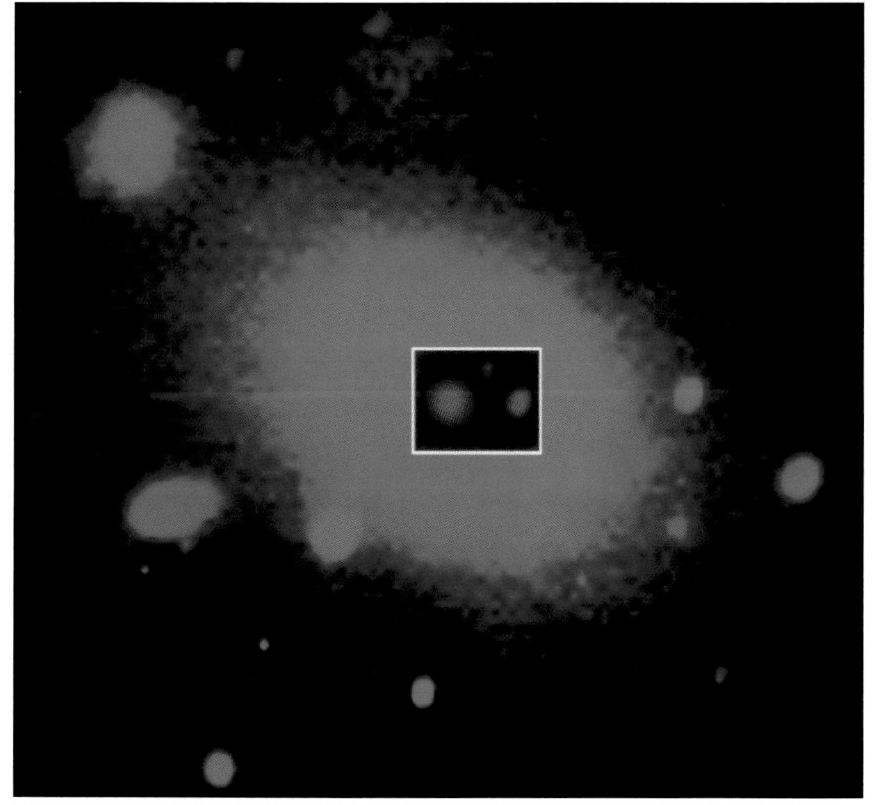

Optical image recorded with a CCD camera on Polaroid Polacolor ER Land Film Type 809.

(on preceding page) Abell 2199, observed at optical wavelengths with a CCD camera, is an example of a galactic cannibalism, a process by which smaller galaxies trapped in the gravitational field of a large neighbor ultimately become part of the larger galaxy. By using the intrinsic large dynamic range of the CCD camera and digital image processing techniques, the galaxy is shown in blue with a "window" revealing that within the core of Abell 2199 are several smaller galaxies that have been captured by Abell 2199. Surrounding the galaxy are numerous smaller galaxies, which may be in the process of being cannibalized.

(below) NGC 3992, an example of a barred spiral galaxy as observed with a CCD camera at optical wavelengths. The spiral arms are made up of young blue stars and star clusters. The inner arms appear to terminate in a bar structure that rotates with the galaxy. Many spiral galaxies (approximately one-third) show a barred structure; the reason for this is still unknown, but it appears to be a stable form to which many galaxies aspire.

(top, right) Sextans A, with its diamond shape and spiral arms, is an unusual example of an irregular galaxy. The brightest dozen stars are in our own galaxy and are not part of Sextans A. Among the faint stars in the galaxy, many have formed into clouds and clusters of hot blue stars. This image was obtained with a CCD camera at the Whipple Observatory.

(bottom, right) Leo I, in an enhanced three-color image (blue-green, red, and infrared) taken with a CCD camera. Leo I is one of a pair of a dwarf elliptical galaxies in the Local Group, a collection of gravitationally bound galaxies including the Milky Way and the Andromeda galaxies. The brightest stars in this galaxy are yellow, while many of the fainter stars resolved in this image are blue. The computer color enhancement enables stars with unusual colors to be easily distinguished and classified.

Optical images recorded with a CCD camera on Polaroid Polacolor ER Land Film Type 809.

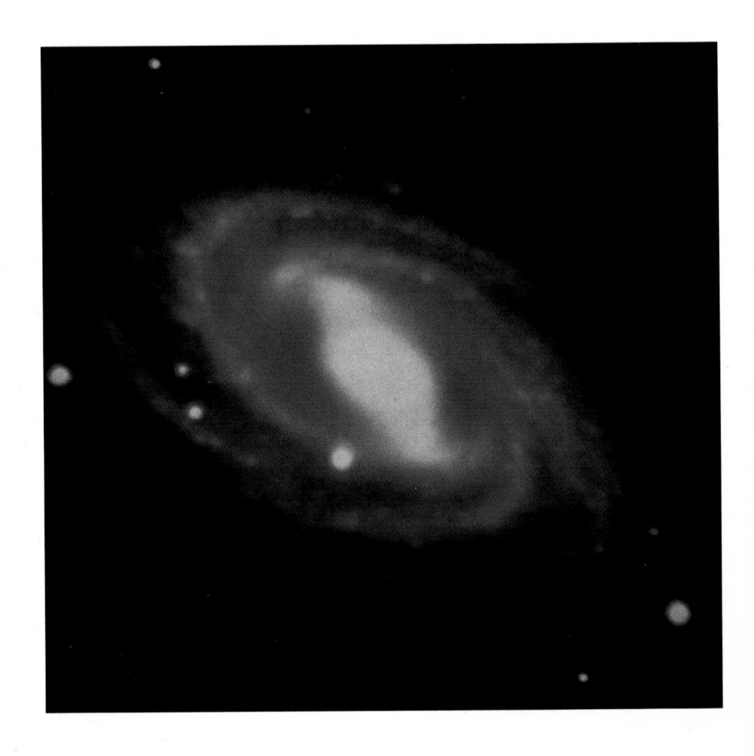

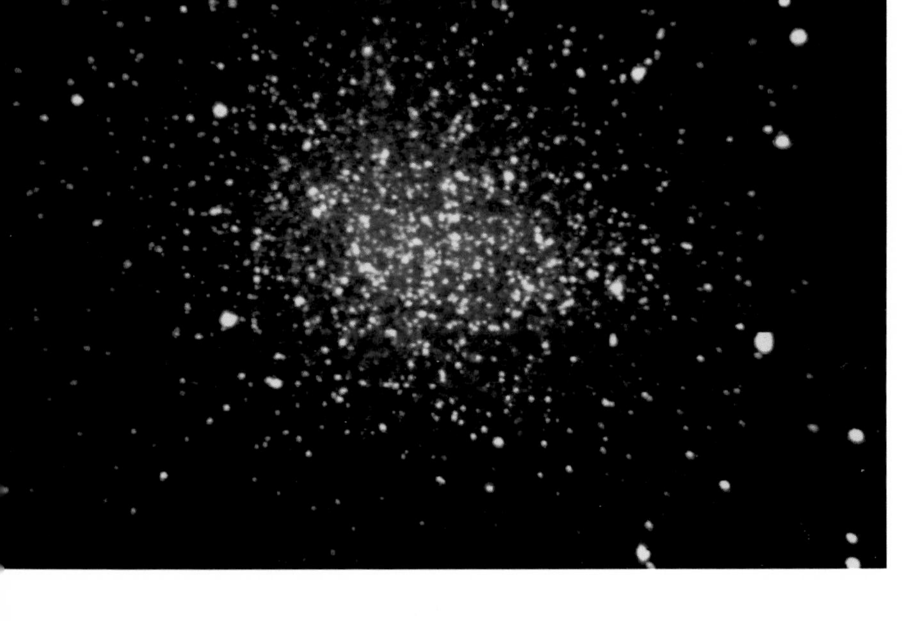

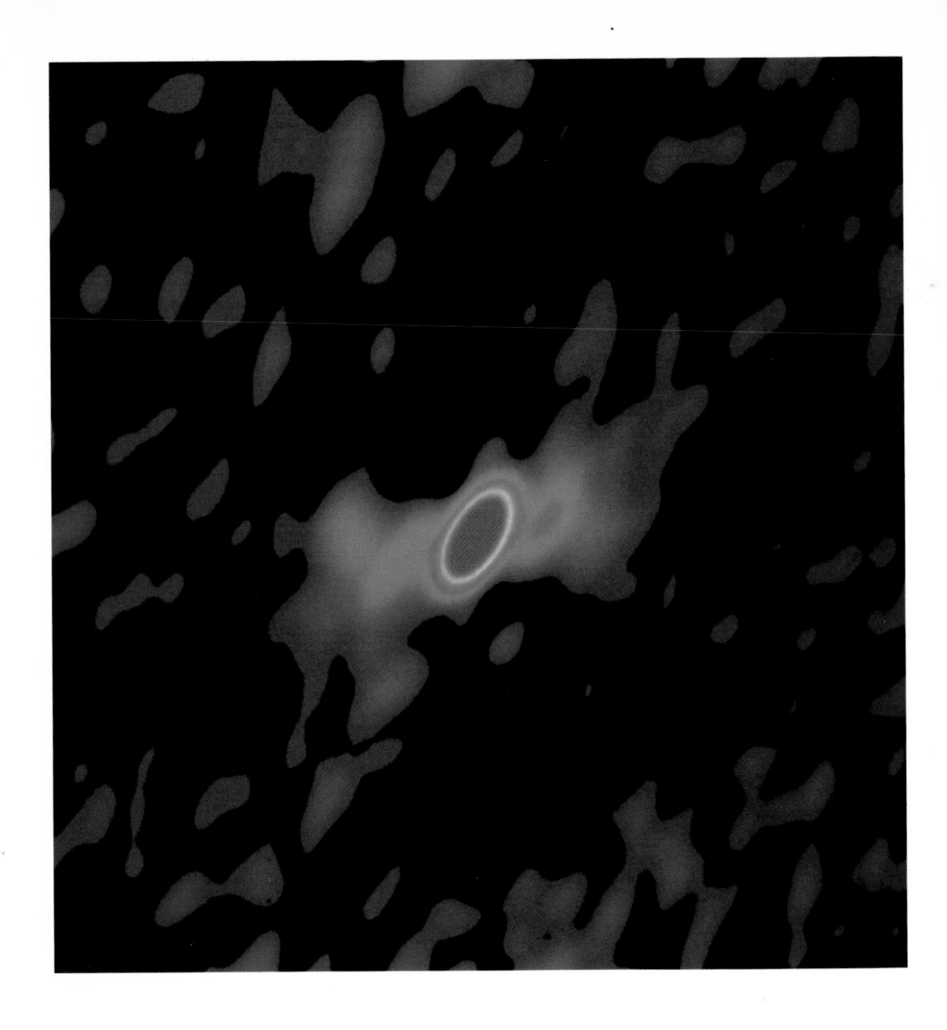

SS433, shown here at radio wavelengths from observations made with the Very Large Array, is one of the most unusual objects in the sky. SS433's peculiar nature became evident in 1978 when it was realized that this object, first observed by optical astronomers, was also visible at radio and x-ray wavelengths. Observations of the spectrum of SS433 have led to the interpretation that it is actually a double star system. One member of the system is an ordinary star while the companion is a much more compact neutron star or possibly a black hole. The pair are orbiting about one another, and the normal star is losing stellar material equal to more than 100 earth masses per year as it is pulled into an accretion disc about the giant companion. As this material accelerates towards the massive companion, it emits a large flux of x-rays. In addition, there are two jets of gas which are being ejected on either side of the disc and which are precessing with a period of about 5½ months much like the wobbling of a top; the spectrum of SS433 shows that the jets are moving at 29% of the speed of light. In this photograph we see the central core of the system with radio jets on either side.

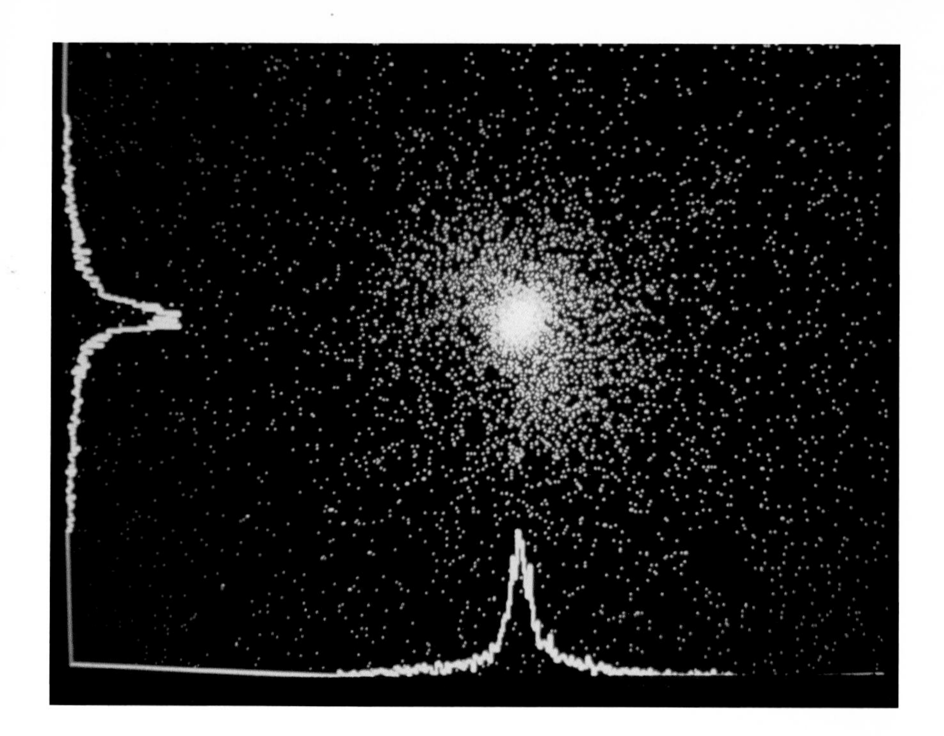

Vela, a supernova remnant, as seen at x-ray wavelengths with the Einstein Observatory. At optical wavelengths, Vela is a beautiful and intricate web of filaments of hydrogen and other gasses, the material ejected during the supernova explosion. The image shown here reveals that the center of this supernova is radiating brightly in x-rays. This central spot, probably a rotating neutron star, has also been observed to be pulsing thirteen times per second at both visible and radio wavelengths. This pulsar is 400 times fainter than the Crab Nebula pulsar at optical wavelengths and 1000 times fainter than the Crab Nebula pulsar at x-ray wavelengths.

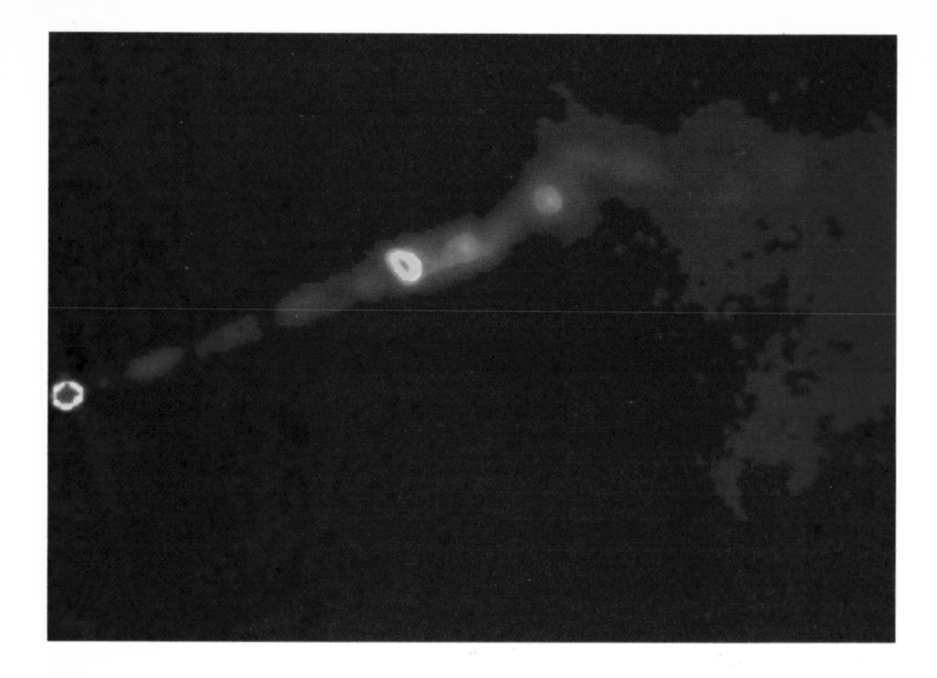

(above) The jet of M87 is here shown as it is observed at radio wavelengths with the VLA. The correlation between optical and radio features is very striking. The center of M87 is enormously dense, but it radiates anomalously little energy at visible wavelengths. It is speculated that the jet, which is being ejected at enormous speeds, is the remnant of stars that, as they race towards a supermassive black hole at the center of the galaxy, release some of their gas. The gas is flung outwards at a high rate of speed by energetic processes near the black hole.

M87 at optical wavelengths *(top, right)* is a giant elliptical galaxy in the Virgo cluster. To make a galaxy with the mass of M87 would take three trillion suns. M87 also has a pronounced jet composed of hot ionized gas that is being shot from the galaxy at a fraction of the speed of light. The jet of this huge galaxy may be ejected along the axis of rotation; there is some evidence for a counterjet as well. Using image processing techniques, the galaxy has been suppressed *(middle)*, leaving only the jet. We can see that the jet is really a sequence of gaseous knots. Finally, the jet is superposed on the galaxy *(bottom, right)*, and it is now clear that the jet extends all the way to the galactic center.

Optical images recorded with a CCD camera on Polaroid Polacolor ER Land Film Type 809.

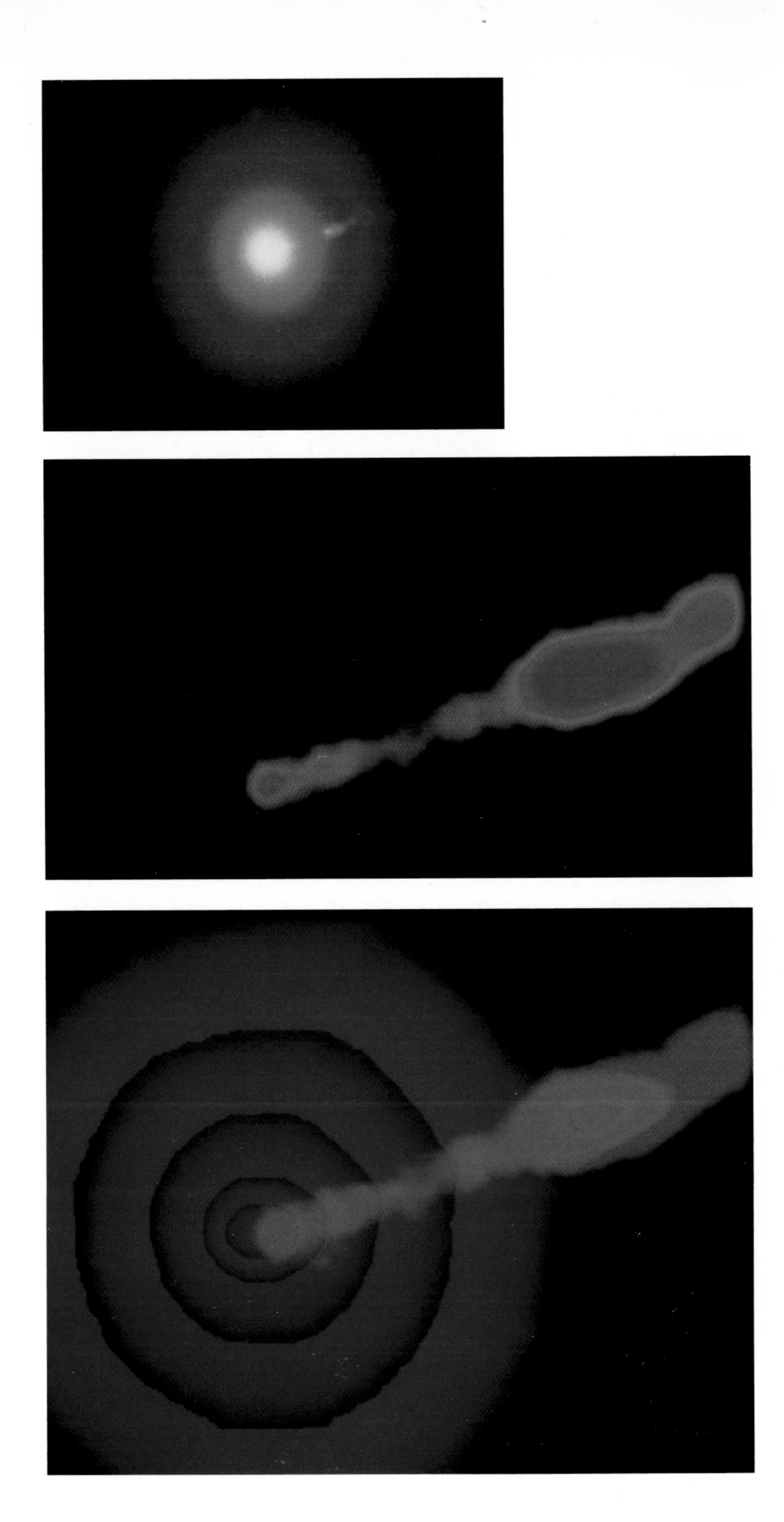

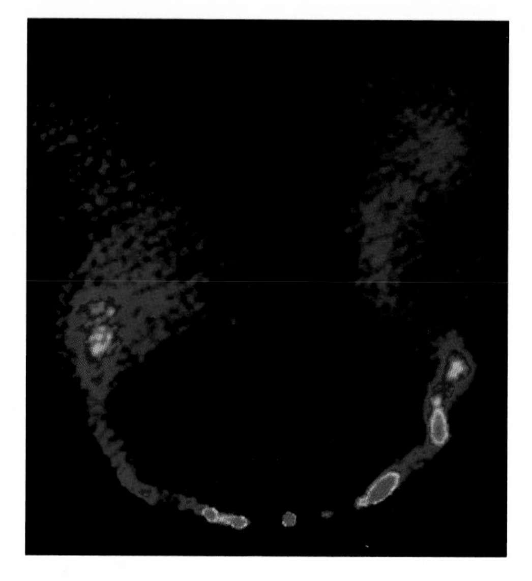

(left) NGC 1265 is a galaxy (here observed at radio wavelengths by the VLA) in the Perseus cluster, which is moving through the intergalactic medium at a velocity of approximately 2000 km/sec. The striking pair of jets are thought to be ionized gas ejected from the galactic nucleus. The jets are bent into the observed horseshoe shape as a result of the movement of the galaxy through the surrounding intergalactic gas.

(below) The Hercules cluster shown here in a three-color CCD visible-wavelength image. Here we see a rich assortment of galaxies, including spirals and ellipticals. Some galactic clusters are tightly bound collections of galaxies whose structure in some ways resembles globular clusters of stars. Other clusters, such as our own Local Group and the Hercules cluster, are loosely bound; the members of such clusters wander in wider, looping orbits. Of particular interest are the apparently colliding spiral galaxies near the center of this image.

Optical image recorded with a CCD camera on Polaroid Polacolor ER Land Film Type 809.

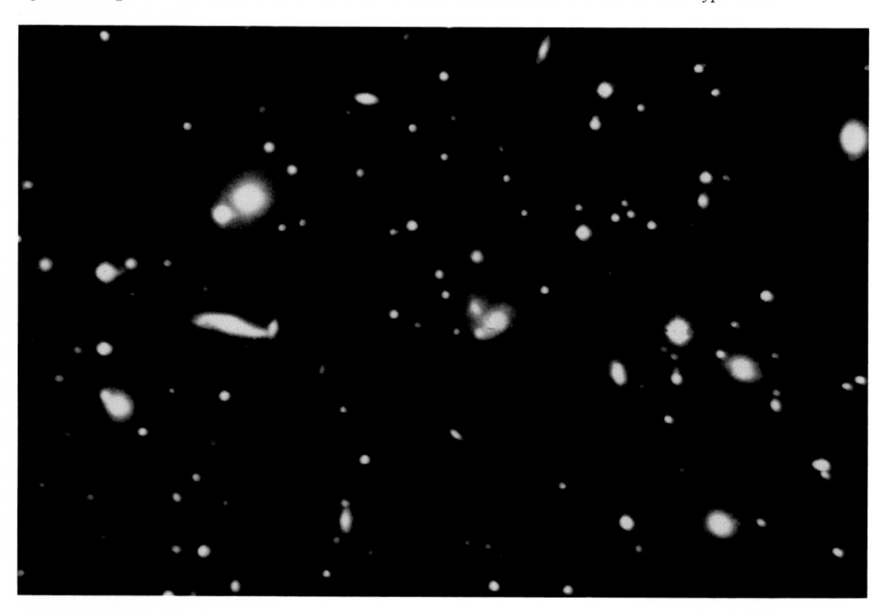

the quasar itself or in intergalactic clouds lying along our line of sight. In the latter case, the clouds should contain very little carbon or other medium-weight elements created in stars, because such gas would never have been inside a galaxy. The gas in such clouds would therefore be a good candidate for the study of helium and deuterium in their true primordial amounts. Such observations, however, must be made at much shorter wavelengths than are accessible to ground-based observatories. With the ultraviolet instruments aboard the *Space Telescope*, we should be able to determine the helium abundances in clouds having large Doppler shifts and the deuterium abundances in clouds of all but the very lowest Doppler shifts.

If the past decade is any guide, important and totally unexpected phenomena will be discovered whenever new wavelength regions are experimentally broached or when sensitivity in existing bands is significantly increased. For extragalactic research, as for galactic studies, the prime unexplored wavelength ranges currently lie in those regimes that are obscured by Earth's atmosphere; the far-infrared region, the gamma-ray domain, and the submillimeter region of the radio spectrum. Are there major astrophysical phenomena yet to be discovered in these last frontiers of the electromagnetic spectrum? Recent progress has taught us that imagination is no limit while predicting what new and exciting findings await us. Nor could Galileo himself have imagined what we "natural philosophers" four centuries hence now know about the Universe beyond the optical domain that he so ably explored with his newly polished ocular.

Cosmic Background Radiation

In 1965, I had just come to Berkeley as a professor, when a friend back east called to report that Penzias and Wilson had discovered a cosmic background radiation at a wavelength of 7 centimeters. Their value for the temperature of the radiation was 3.5 ± 1 degrees. At once, I recalled an unsolved problem in the physics of the interstellar medium – why do interstellar cyanogen (CN) molecules seen in the spectra of distant stars rotate, as shown by optical astronomers in 1941? As stated by Gerhard Herzberg, Nobel Laureate in chemistry, in the 1950 edition of his book, *Spectra of Diatomic Molecules,* "From the intensity ratio of the lines with $K = 0$ and $K = 1$, a rotational temperature follows, which has of course only a very restricted meaning." Little did he realize that the "restricted meaning" to which he referred would turn out to be the temperature of the Universe as a whole, a discovery which earned Penzias and Wilson the Nobel Prize in physics.

I had worked on this problem off and on, and had finally concluded that there must be a previously undetected component of radiation at a wavelength of 2.6 millimeters, whose temperature is about 2.3 Kelvins (or some –270 degrees Celsius). I had not published this result because an important property of CN – its electric dipole moment – which was needed to clinch the argument, had not yet been measured at that time.

It seemed to me that, given the imprecision in each temperature determination, my hypothetical radiation and that observed by Penzias and Wilson were probably one and the same. To prove it, I needed better optical measurements of the rotation of CN molecules in space and a laboratory measurement of the CN dipole moment. Miraculously, both were available only meters from where I sat. It happened that a graduate student in the next office, John Hitchcock, and a well-known astronomer at Lick Observatory, George Herbig, had just obtained some much better data on the rotation of CN.

Moreover, I had just thrown into the wastebasket the galley proof of a paper on comets which mentioned CN; recovering the paper, I found that I was able to infer the CN electric dipole moment from the spectra of comets. I did the necessary calculations which proved that my radiation and Penzias and Wilson's were the same. John Hitchcock and I wrote up the paper, polished it at a nearby cafe over lunch, and sent it off.

Our conclusion, that the spectrum of the background-radiation is identical to that of a blackbody over the 27-fold wavelength interval from 2.6 millimeters to 7 centimeters, has stood the test of time. That this should be so is a strong indication that the radiation is a relic of the early Universe. The unity of science manifested by the relationship between the cosmic microwave background, interstellar molecules, and the spectra of comets, is a source of deep satisfaction to me.

G.B.F.

8 Forces of Nature

The Limits of Knowledge

I have read, by order of the Most Reverend Father, Master of the Sacred Palace, this work, The Assayer; and besides having found here nothing offensive to morality, nor anything which departs from the supernatural truth of our faith, I have remarked in it so many fine considerations pertaining to natural philosophy that I believe our age is to be glorified by future ages not only as the heir of works of past philosophers but as the discoverer of many secrets of nature which they were unable to reveal, thanks to the deep and sound reflections of this author in whose time I count myself fortunate to be born – when the gold of truth is no longer weighed in bulk and with the steelyard, but is assayed with so delicate a balance.

The Imprimatur to Galileo Galilei's
The Assayer, Rome, 1623.

AN AGE-OLD DREAM – understanding all the forces of nature as different aspects of a single fundamental force – has actually been realized to some extent during the past decade. A theory that unifies two of the four known basic forces was successfully developed, thus merging the electromagnetic force that holds together atoms and molecules with the weak nuclear force that governs the decay of radioactive matter. A comprehensive theory of a third type of force, namely the strong nuclear force that binds subatomic particles, has also been developed. Furthermore, concerted efforts aimed at unifying in turn both of these theories are now being made at many laboratories around the globe.

Astronomical observations have played a role in some of these recent developments and will likely play an even greater role in the future, as the pace quickens in our quest to include the fourth known force – gravity – in what seems destined to become a truly unified theory of all forces.

The notion of invisibility extends beyond the unseen objects in interplanetary, interstellar, and intergalactic space. The microscopic domains of elementary particles and the fundamental forces between them are equally invisible – as they act at a level far below that which we strain to see with the highest powered microscopes. We visualize their effects but cannot perceive them directly. We use the tools of technology to infer their properties, but are unable to view them *per se.* Now, it seems, astronomers and physicists are moving toward an intellectual synthesis of the macro-domain of astrophysics and the micro-domain of particle physics. While extending our senses beyond wonder into the realm of understanding, our awe at the working of Nature is renewed.

Energy Sources in the Universe

Newton's law of gravitation, built on Galileo's principles of physics, set the stage for the investigation of the forces of nature that continues today. We now realize that chemical energy, such as that released in the burning of fossil fuels, results from the action of electrical forces among the charged particles within atoms. Holding electrons in orbits around atomic nuclei much as gravity holds planets in their orbits around the Sun, electrical forces release energy whenever an electron drops into a lower orbit.

Magnetic forces also result from electrically charged particles, and so it was natural to try to unify these two – electric and magnetic – forces. In the 1860s, James Clerk Maxwell succeeded in doing just that by developing an "electromagnetic theory". His theory goes on to explain electromagnetic radiation as a wave that sustains itself through a constant interplay between electrical and magnetic forces. Maxwell's ideas form the basis for much of our discussion of the nature of radiation in Chapter 1, though he was largely unaware of the existence of radiation beyond the visible spectrum. By the end of the 19th century, thanks to Newton and Maxwell, both gravitational and electromagnetic forces were well understood at a certain level.

Early in the 20th century, the world of physics experienced a veritable revolution. A series of crucial experiments revealed that the orbits of electrons around atomic nuclei differ qualitatively from those of planets around the Sun. The position of a planet can be predicted precisely by computing the effect of the gravitational force acting on it, but the best we can do with an electron is to predict the probability of its being at various possible positions. The impossibility of doing any better, embodied in Heisenberg's Uncertainty Principle, is an essential feature of what we now call "quantum theory".

In turn, an amalgam of electromagnetic theory and quantum theory was achieved during the 1950s. Called by the tongue-twisting name of "quantum electrodynamics", this theory is now unchallenged in its ability to describe the electromagnetic interactions between subatomic particles such as protons, electrons, and neutrons. A shining goal of contemporary physics is to bring the understanding of all the forces in nature up to the standard of quantum electrodynamics.

Sunlight is electromagnetic radiation, and the form in which the energy of sunlight is stored in plants is chemical energy; both of these forms of energy are embraced by the theory of quantum electrodynamics. But what about the origin of the Sun's energy itself, which is ultimately released as sunlight? Early suggestions included electromagnetic radiation trapped within the Sun, chemical energy stored in its atoms and molecules, and the energy due to the gravitational attraction among all of its atoms. However, none of these forms of energy is adequate to keep the Sun shining for its known age of nearly 5 billion years.

The solution to this problem was reached when physicists discovered a new form of energy during the first part of this century. We now realize that nuclear energy, released for example when the strong nuclear force between hydrogen nuclei (*i.e.*, protons) fuses them into helium nuclei, can keep our Sun shining for many billions of years.

Nuclear interactions occur only at very high temperatures; only then do atomic nuclei have sufficiently high speeds to overcome their mutual electrical repulsion. Thus nuclear forces play a role in astronomy only where matter is extremely hot, as in the interiors of stars or the searing heat of the big bang. In the last few decades, laboratory studies of nuclear reactions have shown that nature actually displays two types of nuclear force, strong and weak; the latter is associated with the unusual neutrino particle discussed earlier in Chapter 3.

Earlier we referred to progress toward unification of the forces of nature. Maxwell showed that electricity and magnetism can be understood as one phenomenon – electromagnetism. In the last two decades, physicists have proved that the weak nuclear force is, along with electromagnetism, part of a more general theory now termed the "electroweak" theory. How does it work? Well, in quantum mechanics, forces between two particles are represented by the exchange of another type of particle called a boson; in effect, the two particles play a rapid game of "catch" using a boson as a ball. In electromagnetism, the boson is a photon – a quantum of electromagnetic energy that always travels at the speed of light. In the new theory, there are four bosons: the photon, the W^+, the W^-, and the Z^0. At temperatures above 10^{15} degrees, these bosons work together in such a way as to make indistinguishable the weak and electromagnetic forces. But at lower temperatures – which includes almost everything we know about – the bosons split into two families: the photon and the other three. The photon carries the electromagnetic force, while the other three bosons carry the weak force.

All of this new physics was developed by the originators of the theory on the basis of a mathematical model and keen insight as to how nature works. In 1983, experimenters in Geneva confirmed the theory by discovering the W and Z bosons in a particle accelerator that was able, for a brief instant, to create a tiny region where the temperature in fact reached some 10^{15} degrees. This success has given physicists new encouragement to search for a unified theory that also includes the strong nuclear force, as explained further below.

Two Puzzles: Solar Neutrinos and Hidden Mass

Neutrinos can penetrate the entire Sun, so weak is the force with which they interact with matter. Detectors placed close to nuclear reactors, which are copious sources of neutrinos, can record only a minute fraction of those emitted. Even so, and despite their elusiveness, the role of neutrinos in astronomical research has become increasingly important.

As noted in Chapter 3, the widely accepted theory of stellar energy generation predicts that large numbers of neutrinos are produced during the fusion of hydrogen into helium in the core of the Sun. But, as also noted in that chapter, researchers have been able to observe only about one-third of the neutrinos predicted to be emitted by the Sun on the basis of the most carefully constructed models of the solar interior. Among the many proposed explanations of this discrepancy is the notion that neutrinos are more complicated particles than heretofore thought possible.

As we saw in Chapter 7, neutrinos might also have some relevance to a comletely different area of astronomical research. In recent years, astronomers have proposed that the "hidden-mass" problem among galaxies might be resolved if the rest mass of neutrinos were not zero, as physicists have until now assumed. Calculations show that the great numbers of neutrinos created in the aftermath of the big bang could supply the hidden mass in galaxy clusters, provided each neutrino has a mass of about 1/10,000th that of an electron.

Thus, two major astronomical problems could conceivably be resolved if neutrinos prove to have properties not previously known. In a series of sweeping proposals, theoreticians have recently argued that both of these problems can be simultaneously solved by merging a pair of current ideas. First, physicists have now constructed a comprehensive theory of the strong nuclear force, called "quantum chromodynamics". This theory, which postulates the existence of quarks of fractional electric charge that make up protons and neutrons, has had considerable success in explaining the results of a wide variety of experiments with subatomic particles, and thus gives us increasing confidence that it correctly describes the strong nuclear force that binds neutrons and protons into atomic nuclei.

Second, spurred by the recent success of the unification of the electromagnetic and weak nuclear forces into the electroweak force, as well as of quantum chromodynamics, physicists are now trying to find an even more general theory that incorporates the electroweak and strong nuclear forces in a "grand unified theory", dubbed GUT for short. While there are currently several different versions of this GUT (no one of which has yet clearly emerged as the most promising), some predict that the three different types of neutrinos recognized in particle physics experiments – the e, mu, and tau neutrinos – actually have finite rest masses, and as a consequence can change readily into one another.

If proven experimentally, even a small rest mass for neutrinos could help solve the hidden-mass problem in galaxy clusters. And, because e-neutrinos emitted by the Sun would be expected to change into one of the other types during the eight-minute flight time between Sun and Earth, and since the underground solar neutrino detector described in Chapter 3 is sensitive only to e-neutrinos, the factor-of-3 discrepancy would be thereby explained.

Thus, two of the most formidable problems in contemporary astrophysics might depend critically on the properties of the elusive neutrino particles. A major problem in solar physics could be solved if neutrinos change among their three types, but such transformations can occur only if neutrinos have a finite mass; however, if this is true, we can potentially solve a major problem in extragalactic astronomy. Actually, the stakes may be even higher, for a finite neutrino rest mass would increase the average density in the cosmos, and thus have an important bearing on the ultimate fate of the Universe.

The First Few Minutes of Existence

Although astronomical data now available favor a big-bang cosmology, we cannot yet consider the big-bang model as conclusively proven. Thus, it is of the greatest importance to test its predictions however possible.

The big-bang model predicts that the cosmic microwave background originated as high-temperature radiation in the first few minutes of time. As the Universe expanded, according to this idea, the radiation cooled to its currently observed temperature, about 2.7 degrees above absolute zero (or some −270 degrees Celsius). When the Universe was about 0.0001 second old, its temperature was on the order of a trillion degrees – so hot that the radiation present must have created about a hundred million proton-antiproton pairs for every proton now observed in the Universe. As time passed, these pairs annihilated, leaving behind an excess of protons over antiprotons. Had this excess not existed, the number of protons in the present Universe would have been vastly smaller, and there would not have been sufficient matter in the Universe to form galaxies, stars, planets, and life.

What caused the slight excess of matter over antimatter implied by this big-bang scenario? Until recently, astrophysicists had regarded this excess as a fact as inexplicable as the existence of the Universe itself. Now, we realize that the grand unified theories are likely to provide an explanation: the decay of very heavy particles called X-bosons within the first 10^{-35} second of the history of the Universe created slightly greater numbers of protons than antiprotons. This prediction can be tested in a straightforward way, for if protons can be created they can also be destroyed. Using the grand unified theories, we can estimate the lifetime, or mean life expectancy, of the proton; it turns out to be more than a hundred billion times the age of the Universe. Certainly we need not be concerned that the matter in the Universe will soon evaporate. On the other hand, quantum theory states that, just as people die unpredictably, every proton in the Universe is in danger of decaying at any moment. The enormous value of the life expectancy means that the probability of decay in any given time span is small. Since water is an abundant source of protons, we can predict that on average one proton should decay per year in each ton of water. Experiments are now in progress to detect such events in huge quantities of water stored in tanks in deep mines, where the water can be insulated from the spurious effects engendered by cosmic rays reaching the Earth from outer space.

Inflation: The First Instants of Time

If grand unified theories are correct, they have another important implication for the Universe in the first instants of its existence. Try to contemplate times as short as 10^{-39} second after the creation, when the temperature was some 10^{30} degrees. At that time, only one type of force other than gravitation operated – the grand unified force. From the theory of such a force, we find that the matter of the Universe exerted a very high pressure that pushed outward in all directions. According to the general theory of relativity, which describes the effect of gravitational forces in the Universe, the Universe must have reacted to this pressure by expanding in a regular manner: for every factor of 100 increase in time, the size of the Universe increased by a factor of 10. The temperature dropped as the Universe expanded, in proportion to the reciprocal of the size of the Universe. Thus, as time advanced from 10^{-39} second, the Universe expanded another order of magnitude, and the temperature fell to 10^{28} degrees.

Suddenly there was a dramatic change. The Universe exploded, increasing its size by the huge factor of 10^{20} in a mere 10^{-35} second. Why? Grand unified theory postulates that the temperature of 10^{28} degrees is extremely critical, so that dramatic changes were bound to occur once that value was reached. This is the temperature at which particles carrying the grand unified force (the so-called X-bosons) could no longer be produced, because the energy embodied in each such particle was no longer available owing to the falling temperature. Therefore, as the temperature approached 10^{28} degrees, the X-bosons disappeared, and instead of exerting an outward pressure, the matter in the Universe suddenly began to exert an inward tension. The tensile strength of the matter was unimaginably huge – 10^{96} times that of the best steel. Intuitively we might imagine that such an enormous inward force would stop the Universe from expanding, or at least slow it down. But general relativity teaches us differently; the actual effect of the sudden change from outward pressure to inward tension was a rapid acceleration in the rate of expansion. This period of rapid expansion has been called "inflation". If this speculative theory is correct, the Universe inflated some 10^{20} times or more in a mere 10^{-35} second.

We might well wonder what happened to the energy that must have developed as the Universe expanded rapidly against its restraining tension force. The answer is the same as for a steel cable being stretched: the energy is stored in the matter of the Universe, which at that time was in a strange state known as a false vacuum. Huge amounts of energy were stored in the false vacuum, but this state could not persist forever. Just as with a stretched steel cable, the matter of the Universe had to "break". Break it did, suddenly releasing the stored energy into heat, and raising the temperature back to 10^{28} degrees or thereabouts.

At the conclusion of inflation, the X-bosons had disappeared forever, and with them the grand unified force. In its place were the electroweak and strong nuclear forces operating around us today. With the new forces, the matter once again exerted an outward pressure, and the Universe returned to its more leisurely expansion.

What, then, do we make of inflation? Did it leave any traces for which astronomers might search today? The answer is a resounding yes. For one thing, inflation had the consequence of putting the Universe into a state precariously balanced between infinite expansion and ultimate recollapse. Recall from Chapter 7 that for this to happen, the density of matter must equal a critical value just sufficient for its gravitational effect to retard the rate of expansion. Astronomers have demonstrated, as explained earlier, that the amount of normal matter (protons, neutrons, electrons) is only about 10 percent of the critical value. If the concept of inflation is correct, then 90 percent of the matter in the Universe is not ordinary, but in some unorthodox form such as black holes, massive neutrinos, or particles not yet known to physics. In other words, most of the matter in the Universe must be invisible. It is striking that, as we have explained earlier, astronomers have uncovered evidence for hidden matter in the outer parts of galaxies and in clusters of galaxies. Is this the matter that inflation predicts?

Very recently, particle physicists have deduced another consequence of inflation. They reason that extremely small-scale fluctuations in the matter density before inflation (which are the inevitable consequence of quantum theory) would be stretched by inflation to an extent now embraced by whole galaxies and clusters of galaxies. Recall from Chapter 5, when we addressed the issue of galaxy formation, it was just such fluctuations, laid down in the early Universe, that theoretical astrophysicists require in order to understand how galaxies originate. Thus, a mind-boggling concept emerges: The vast agglomerations of matter we see today as galaxies, clusters of galaxies, and superclusters of galaxies, are remnants of quantum fluctuations that occurred when the Universe was only 10^{-35} second old. If the notion of inflation is correct, particle physics and cosmology have been joined in an intellectual synthesis that no one could have predicted. With their largest telescopes probing the deepest reaches of space, astronomers may actually be testing whether grand unification is a correct theory of particle physics.

Closing In On Creation

Gravity keeps us on Earth, binds Earth to the Sun, keeps the Sun in orbit in the Galaxy, and slows the expansion of the Universe. Newton described it as a force, while Einstein, in his Theory of General Relativity, interpreted gravitational forces in terms of the curvature of space-time. Decades of experiments have proved Einstein's ideas to be an accurate representation of reality. General Relativity is thus crucial if we are to understand cosmic systems such as neutron stars, black holes, and the expanding Universe.

The Theory of General Relativity predicts a new form of energy called gravitational radiation. This radiation is expected to propagate through space in the form of waves that will accelerate any material object they encounter, in much the same way that electromagnetic waves propagate through space and accelerate any electrically charged object they encounter. Although this radiation has not yet been directly detected, several research groups are now building equipment ("gravitational telescopes") to measure the accelerations due to passing gravity waves; this equipment promises to be thousands of times more sensitive than anything currently available.

What events in the Universe might emit gravitational radiation? Virtually every object to some extent, but with an efficiency far too low to be currently detectable. We cannot easily calculate the amount of gravitational radiation with precision, but estimates so far imply that our best chance for detection lies with catastrophic events like the collapse of a star to form a neutron star or stellar black hole or the collapse of the core of a galaxy to form a supermassive black hole. A key requirement is that the collapse must occur at nearly the speed of light, and must not be spherically symmetric. The latter property reflects the fact that if the collapse is symmetric, any gravitational radiation produced remains near the object, and does not propagate to large distances where it can be observed. It is this spherically asymmetric aspect which no one can yet quite fathom that makes our estimates in this field so crude. Nonetheless, even if our efforts are only approximately correct, we might be able to detect gravitational radiation before the end of the century from supernova explosions in the Virgo cluster of galaxies. If a supernova unpredictably bursts forth within our Galaxy, we should be able to detect its gravitational radiation with ease. (Gravitational radiation has already been inferred to exist, as the orbit of a pulsar in a binary-star system has been proved to be slowly shrinking; the corresponding energy loss is just that calculated for the emission of gravitational waves from an orbiting neutron star.)

Efforts to develop a quantum theory of gravitation have not yet succeeded, but we have ample reason to expect that quantum effects should occur near black holes, where space-time is so highly curved that the microscopic domain is fully evident. The quantum theory of elementary particles predicts that even in a region of space containing no matter or energy (a vacuum), particle-antiparticle pairs are constantly created and annihilated within an interval of time too short to observe. This effect at first seems impossible. After all, how can a particle appear out of nothing? The answer is that no laws are violated because the particle is annihilated by its corresponding antiparticle before either one can be observed. Furthermore, for such events *not* to happen would violate the laws of quantum theory, which cites, via the Heisenberg Uncertainty Principle, the impossibility of determining *exactly* how much energy or mass is present in any system. Hence, fluctuations of energy content must occur even when the average energy present is zero. When this pair creation occurs near a black hole, one member of the pair could conceivably fall into the black hole before the pair has a chance to annihilate. The other member then leaves the scene, making the black hole appear to the outside world as a source of radiation. In this way, black holes are theorized to evaporate slowly while gradually losing energy.

Up to this point, the black holes we have discussed were either of stellar mass or supermassive, having formed by the collapse of stars or the cores of galaxies. Yet it seems possible that black holes of all sizes might have originated in the extraordinary conditions of the big bang. Any such "primordial black holes" having masses of about 10^{15} grams – about the mass of a small mountain on Earth – would be evaporating at just about the current time in the history of the Universe, giving rise at the end of their lives to bursts of gamma-ray radiation. Such radiation from evaporating black holes has been searched for, though none has yet been found. We thereby conclude that, if primordial black holes do exist with masses less than that of mountains, they cannot make up a significant fraction of the mass of the Universe.

The theory of black-hole evaporation depends on the quantum nature of strong nuclear forces, but not on the quantum nature of gravitation. To be sure, no one has yet proposed a convincing theory of gravitation that incorporates quantum principles. Even so, we can safely conjecture that quantum effects must become important whenever the radius of curvature of spacetime becomes less than 10^{-33} centimeter (a distance known as the "Planck length"). Briefly, the argument is this: According to relativity theory, the radius of curvature of spacetime is proportional to the mass enclosed. On the other hand, the Heisenberg Principle states that no particle can be located with a precision better than a distance which is inversely proportional to the mass. For ordinary objects like stars, the first distance is vastly larger than the second, and there is no contradiction. As the mass becomes smaller, however, the radius of curvature decreases and the distance uncertainty increases until they finally equal each other at a mass of roughly 10^{-5} gram, at which mass both the radius of curvature and the distance uncertainty equal 10^{-33} centimeter, the Planck length. For such masses, the distance uncertainty and the radius of curvature are equal, so relativity theory (which does not incorporate the uncertainty principle) is no longer an adequate description of nature. The universal radius of curvature equalled the Planck length some 10^{-43} second after the big bang, at which time the temperature was about 10^{32} degrees Celsius. Owing to a lack of theory incorporating both general relativity and quantum effects, we simply cannot describe the Universe at this time – or any earlier time. To be able to do so, we need a quantum theory of gravitation.

Because the temperature at which the grand unified theories synthesize the electroweak and strong nuclear forces, 10^{28} degrees, is of the same general magnitude as 10^{32} degrees, the temperature at which quantum gravity becomes important, some physicists speculate that we should be able to include the gravitational force in the theory as well, thus creating a "super-grand unified theory" incorporating all four known forces in nature. The resulting super-grand force would operate as a single force at energies only slightly higher than those relevant to the currently fashionable theories of grand unification. Only at smaller energies (or at times after 10^{-43} second), would the more familiar four forces begin to manifest themselves distinctly, though in reality all four are merely different aspects of the single, fundamental, super-grand force that existed at (or near) creation. Some physicists hope that such a super-grand unified theory will yield, almost as a by-product, the correct theory of quantum gravitation. But, because vigorous attempts toward this end have thus far met with little success, the development of such a theory is widely considered to be the currently ultimate challenge to physics.

The notion of force, as a law governing matter once fashioned, fails to account for the process of creation itself. It is possible, as astrophysicists push the frontiers of time back to the moment of cosmic creation, that we shall recognize the existence of the Universe as a consequence of the nature of the fundamental force? Is it possible that the potential existence of the world somehow calls it into existence? Such questions, once believed outside the realm of science, are now arising in scientific thought.

Intergalactic Hydrogen

When I was a graduate student at Princeton in the early 1950s, I learned of the discovery of the 21-centimeter line of interstellar hydrogen in our Galaxy. Its emission had been captured by an antenna and recorded.

As a budding astrophysicist, I knew that any line which could be seen in emission could also appear in absorption under the right conditions; that is, radiation at 21-centimeter wavelength from a distant source would be absorbed by any intervening hydrogen atoms which were colder than the source. I asked for observing time on a radio telescope to test my calculations, but another group stumbled upon the effect before I got my time. So I began thinking about other situations where the effect might be useful.

At the time there was considerable interest – and there still is – in the question of whether the Universe has enough matter to cause it to gravitationally collapse upon itself some time in the (hopefully) distant future. I reasoned that even very rarefied hydrogen gas between the galaxies could contribute significantly, and calculated that if it were in the form of neutral atoms, it could be detected.

Thanks to David Heeschen, the acting director of the newly formed National Radio Astronomy Observatory, I got time on an 85-foot radio telescope to make the required observations. I saw no effect, and concluded that if enough hydrogen were there to be significant, it must be ionized, and thus unable to absorb 21-centimeter radiation.

With the discovery of quasars at great distances, James Gunn and Bruce Peterson were able to repeat the test using atomic transitions which are far more sensitive. Again, negative results.

It wasn't until much later that researchers discovered that hydrogen atoms are in fact present in intergalactic space. The atoms are located in discrete clouds rather than being distributed uniformly along the line of sight, as we had assumed, so one has to look at exactly the correct wavelengths to see them. The numbers of hydrogen atoms are far too small to be significant for the future of the Universe, but the very existence of clouds is interesting because it suggests that there may be much more hydrogen between the clouds in ionized form. However, the best theoretical models of such atomic clouds yield amounts of ionized gas between them which are not important for the future of the Universe either.

Astronomy is not an experimental science but an observational one. We are stuck with the meager amounts of information the Universe gives us, and there are perfectly good questions which can be asked which may never be answered.

G.B.F.

Epilogue

You see then, studious reader, how the subtle mind of Galileo, in my opinion the first philosopher of the day, uses this telescope of ours like a sort of ladder, scales the furthest and loftiest walls of the visible world, surveys all things with his own eyes, and, from the position he has gained, darts the glances of his most acute intellect upon these petty abodes of ours – the planetary spheres I mean, – and compares with keenest reasoning the distant with the near, the lofty with the deep.

From the preface of *Dioptrics,*
by Johannes Kepler, Augsburg, 1611.

NATURE OFFERS NO GREATER SPLENDOR than the starry sky on a clear, dark night. Silent, timeless, jeweled with the constellations of ancient myth and legend, the night sky has inspired wonder throughout the ages – a wonder that leads our imaginations far from the confines of Earth and the pace of present day, out into boundless space and cosmic time itself.

Astronomy, born in response to that wonder, is sustained by two of the most fundamental traits of human nature: the need to explore and the need to understand. Through the interplay of curiosity, discovery, and analysis – the keys to exploration and understanding – answers to questions about the Universe have been sought since the earliest times, for astronomy is the oldest of the sciences. Yet, not since its beginnings has astronomy been more vigorous or exciting than it is today.

Indeed, we are at the dawn of a new age in space science. Astronomy no longer evokes visions of plodding intellectuals peering through long telescope tubes. Nor does the cosmos any longer refer to that seemingly inactive, immutable regime captured visually by occasionally gazing at the nighttime sky. Modern astrophysics now deciphers a more vibrant, evolving Universe – one in which stars emerge and perish like living things, galaxies spew forth vast quantities of energy, and life itself is understood as a natural consequence of the evolution of matter. Yet, amid the cosmic symphony of visible and invisible matter strewn across the Universe, we humans seemingly play no special role. The rock called Earth is merely a platform on which to develop new technologies and sciences, all of which tend to reinforce the magnificent mediocrity of human life in the Universe.

New discoveries always not only advance knowledge, but also raise new questions. Astrophysicists will encounter many new problems in the decades ahead, but this should neither dismay nor frustrate us, for this is precisely how science operates. Each discovery adds to our storehouse of information, generating a host of questions that lead in turn to more discoveries, and so on, resulting in a rich acceleration of basic knowledge.

Through modern astronomical research, we now realize that we are connected to distant space and time not only by our imaginations but also through a common cosmic heritage: most of the chemical elements comprising our bodies were created billions of years ago in the hot interiors of remote and long-vanished stars. Their hydrogen and helium fuel finally spent, these giant stars met death in cataclysmic supernova explosions, scattering afar the atoms of heavy elements synthesized deep within their cores. Eventually this matter collected into clouds of gas in interstellar space; these, in turn, slowly collapsed to give birth to a new generation of stars. In this way, the Sun and its complement of planets were formed nearly five billion years ago. Drawing upon the matter gathered from the debris of its stellar ancestors, the planet Earth provided the conditions that ultimately gave rise to life. Thus, like every object in our Solar System, each living creature on Earth embodies atoms from distant realms of our Galaxy and from a past far more remote than the beginnings of human evolution.

Although ours is the only planetary system we know for sure, others may surround many of the hundreds of billions of stars in our Galaxy. Elsewhere in the Universe, beings with an intelligence surpassing our own may also at this moment gaze in wonder at the nighttime sky, impelled by even more powerful imaginations. If such beings exist – possibly even communicating across the vast expanses of interstellar space – they, too, must share our cosmic heritage.

Emerging largely from our studies of the invisible Universe, this recognition of our cosmic heritage is a relatively recent achievement in astronomy. However, it is but one of many such insights that our generation alone has been privileged to attain. Indeed, our descendants will likely regard our generation as the one that broached the electromagnetic spectrum beyond visible light, thus not only providing a whole new glimpse of our richly endowed Universe, but perhaps more significantly recognizing life's integral role in the cosmos.

In all of history, there have been only two periods in which our perception of the Universe has been so revolutionized within a single human lifetime. The first occurred nearly four centuries ago at the time of Galileo; the second is now under way.

List of Appendices

Appendix A

Space Telescope

Scheduled to be launched by NASA's *Space Shuttle* in 1986, the *Space Telescope (ST)* will provide the first permanent optical observatory in space. Taking full advantage of its location above the distortions, glow, and absorptions of Earth's atmosphere, the *ST's* 2.4-meter mirror will be able to observe cosmic objects a hundred times fainter, and resolve details about ten times finer (0.05 arc second) than can the best optical telescopes now operating on the ground; it will also be able to detect radiation outside the conventional optical spectrum, extending its sensitivity into parts of the ultraviolet and infrared bands of the electromagnetic spectrum. As such, the *ST* will be the first orbiting optical-ultraviolet telescope large enough to undertake studies of extragalactic objects at the limits of the observable Universe. In short, the *ST* is destined to represent one of the most momentous advances in astronomical instrumentation since Galileo's first telescope.

A sketch of *ST's* vital components is shown below.

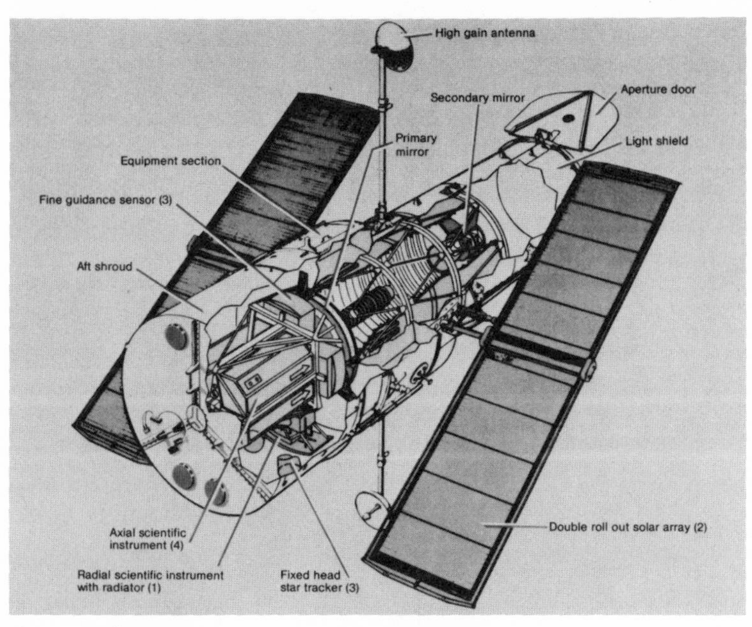

Courtesy of NASA

Appendix B

Submillimeter-Wave Radio Telescope

The interval from 0.8 to 0.3 millimeter wavelength comprises one of the last electromagnetic windows through which astronomical observations can still be made from Earth's surface. Recent advances in the design and construction of ultraprecise antennas and sensitive radio receivers make possible an inexpensive but extremely powerful *Submillimeter-Wave Radio Telescope* that could, from a high-altitude site such as Mauna Kea, observe a significant portion of this largely unexplored band. Such an instrument will allow detailed study of this rich region of the spectrum with an angular resolution as fine as 8.5 arc seconds, which is a factor of two better than that of any existing millimeter-wave antenna.

The fascinating problems of star formation, stellar mass loss, and galactic structure stand at the center of many of the controversial issues of astrophysics today. The important information that this instrument can provide on these topics elevates it to an importance much greater than would be suggested by its modest cost. A sketch of its central components follows. The telescope is in the center, with the reflector directed to the left. The structure around it (peeled away to reveal the telescope) is a dome to protect the telescope from the wind and to reduce temperature excursions.

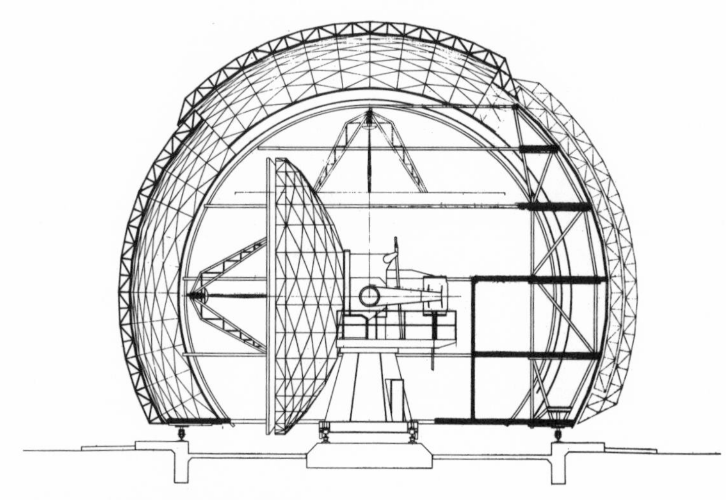

Courtesy of Dr. Robert Leighton, California Institute of Technology

Shuttle Infrared Telescope Facility

The *Shuttle Infrared Telescope Facility (SIRTF)* is scheduled to become the first major infrared observatory in space. In Earth orbit and thus free of the limitations imposed by the Earth's atmosphere, the *SIRTF* will permit investigations throughout the vast range of infrared wavelengths from 2 to 700 microns. Having a 0.85-meter aperture and supporting instruments cooled to temperatures near absolute zero, this shuttle-launched system will be, for certain objectives, up to a thousand times more sensitive than any previously built ground-based or airborne infrared telescope; the gain in sensitivity is so large that it seems a virtual certainty to advance greatly our cosmic database in the infrared part of the electromagnetic spectrum.

Because of its relatively wide field of view, the *SIRTF* will be able to undertake efficient surveys of infrared sources that will help optimize the observing programs of larger instruments, such as the *New Technology Telescope* (*cf.*, Appendix F), the *Very Long Baseline Array* (*cf.*, Appendix H), and the *Large Deployable Reflector* (*cf.*, Appendix G), all of which have narrower fields of view. The diagram below illustrates the central features of this versatile *SIRTF*.

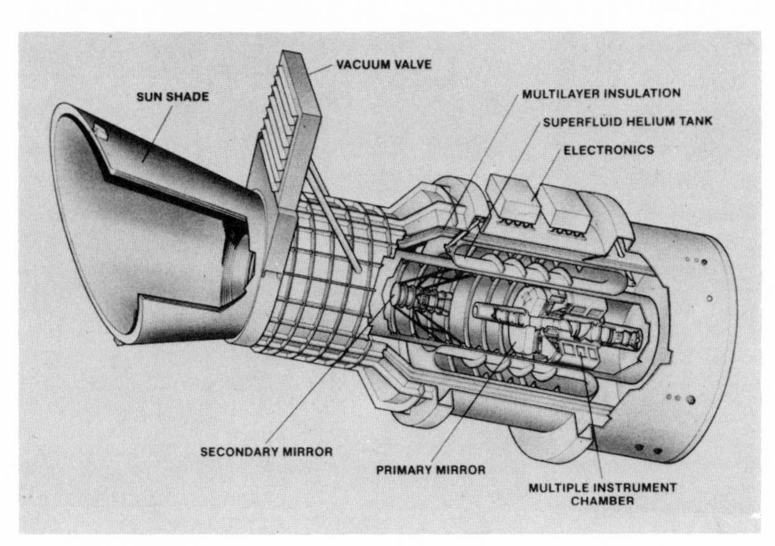

Courtesy of NASA

Appendix D

Advanced X-Ray Astrophysics Facility

The *Advanced X-Ray Astrophysics Facility (AXAF)* is the key component of our future program in x-ray astronomy. As part of an effort to obtain x-ray images of the Universe comparable in depth and detail with those of the most advanced optical and radio telescopes, *AXAF* has been designated as the highest priority new project in United States astronomy for the 1980s. Using a grazing-incidence telescope of 1.2-meter aperture that gently guides the high-energy photons to a focus, this permanently orbiting national facility will be able to form images with 0.5-arc-second resolution. *AXAF* will achieve a hundred-fold increase in sensitivity over the *Einstein Observatory* which was, until its batteries expired in 1981, the most sophisticated x-ray orbiting observatory ever flown. In addition, instruments aboard *AXAF* will provide a range of capabilities of measuring x-ray spectra that will far surpass the sensitivity and resolution embodied in any previous x-ray mission.

All the technology required for *AXAF* is within the current state of the art. In effect, the *Einstein Observatory* served as a prototype that demonstrated the feasibility of all the basic concepts of x-ray telescopy involved in *AXAF.* The sketch below, courtesy of NASA's Marshall Space Flight Center, illustrates *AXAF* after deployment by the *Space Shuttle* sometime in the late 1980s.

Courtesy of NASA

Appendix E

Advanced Solar Observatory

An ensemble of telescopes in orbit, the *Advanced Solar Observatory (ASO)* is expected to be the single most important observational facility for solar studies over the next decade and beyond. Having high angular resolution – 0.1 arc second, or about 70 kilometers on the Sun – the *ASO* will be able to observe the solar atmosphere and its extensions into space over the full range of temperatures present (6000 to more than 20 million degrees). It will be particularly adept at addressing questions relating to solar activity.

Assembled in orbit over a period of years beginning in the late 1980s or early 1990s, the *ASO* will ultimately consist of four groups of instruments. A High-Resolution Telescope Cluster will house a 1.3-meter Solar Optical Telescope operating at wavelengths between 1100 and 10,000 Angstroms, a 0.9-meter Extreme Ultraviolet Telescope (500 to 1200 Angstroms), a 0.4-meter X-Ray/Extreme-Ultraviolet Telescope (100 to 500 Angstroms), and a 0.8-meter X-Ray Telescope (2 to 100 Angstroms). Each of these instruments will possess extensive spectroscopic capability, allowing the simultaneous study of spectral features, temperature and density structures, velocity fields, abundances and other physical and chemical properties in the many complex domains of the solar plasma.

The figure below, courtesy of NASA's Marshall Space Flight Center, depicts two of the *ASO*'s major components being operated aboard the

Courtesy of NASA

Space Shuttle. The canister to the right houses the above-described cluster of high-resolution telescopes, while the 50-meter boom to the left supports the long focal length requirements of the Pinhole/Occulter Facility. This latter instrument is designed to pass an occulting mask (at the end of the boom) over an array of detectors and telescopes; the mask contains a variety of pinholes or small apertures that should enable astronomers to produce x-ray images of impulsive solar flares with an angular resolution of 0.2 arc second. The mask will also contain an occulting disk, or "artificial moon", that will simulate a solar eclipse, allowing the first visible, ultraviolet, and x-ray coronal observations with a large-aperture telescope.

The other two major components of the *ASO* include a High-Energy Facility capable of making x-ray and gamma-ray observations not requiring high angular resolution, and a Low-Frequency Radio Facility that will principally study particle acceleration in the solar atmosphere.

According to current plans, once the majority of the component instruments of the *ASO* are developed and tested on the *Space Shuttle,* they will be assembled on a space platform or station and will operate throughout the 1990s as a national solar space observatory, much as will the *Space Telescope* (cf., Appendix A) and the *Advanced X-Ray Astrophysics Facility* (*cf.,* Appendix D).

Appendix F

New Technology Telescope

A giant ground-based optical-infrared telescope, the *New Technology Telescope (NTT)*, will greatly outperform all current telescopes, for *NTT*'s proposed collection area will not only dwarf the largest optical mirror (Russia's 6-meter reflector), but will double the collecting area now provided only by the simultaneous use of all twenty of the largest telescopes in the world. *NTT*'s suggested aperture size of 15 meters results from two prime considerations; First, 15 meters is about the largest aperture that can be successfully exploited on the ground to achieve the highest possible resolution available below the atmosphere and, second, 15 meters in the largest telescope that can confidently be built at this time, based on recent breakthroughs in manufacturing large optical mirrors, mounting them in a highly controlled way, and directing them with precision toward any desired object.

The principal justification for building such a 15-meter class telescope for optical and infrared astronomy is its unequalled capacity for studying the spectra of faint objects at wavelengths between 3000 and 200,000 Angstroms. The order-of-magnitude improvement that this device should grant in light-gathering capacity will roughly mimic the relative improvement brought by Galileo's telescope over the naked eye. Accordingly, the *NTT* is likely to have applications to a very wide range of scientific problems.

One possible design for the *NTT* utilizes the design of the successful *Multiple Mirror Telescope (MMT)* now operating on Mount Hopkins in Arizona. By having six 6-meter mirrors on a common mount feeding a common focus, the *MMT*-inspired design achieves an effective aperture of 15 meters. This innovative concept uses well-established technology. However, in order to successfully bring the beams from each individual mirror to a common focus (to avoid blurring the image), innovative technology is required to ensure that each mirror is rigid and lightweight.

The sketch is by Lola Judith Chaisson, based on an early design by the Kitt Peak National Observatory.

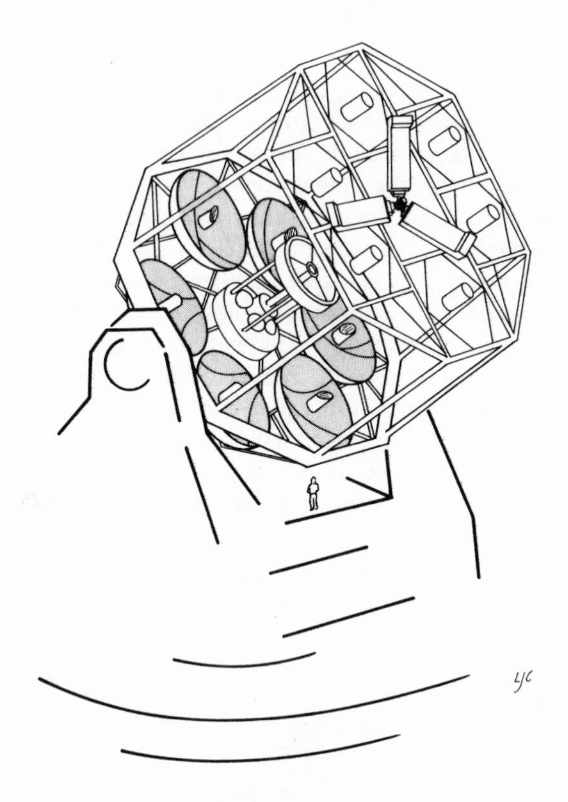

Appendix G

Large Deployable Reflector

Proposed to be launched into Earth orbit, a 10-meter-class *Large Deployable Reflector (LDR)* would be designed to operate in the infrared and submillimeter domains between 0.03 and 1 millimeter wavelength, an electromagnetic band in which atmospheric absorption very effectively blocks ground-based observing. Such a national facility could be deployed with extensive in-orbit construction by astronauts, and would have a long orbital lifetime during which instruments could be changed or upgraded. The *LDR* would be specifically designed to complement the ground-based *New Technology Telescope* (*cf.*, Appendix F) by extending the latter's powerful capabilities to these longer wavelengths; together these two new instruments would guarantee high spectral and angular resolution throughout most of the infrared spectrum. Owing to its lack of cryogenic cooling, the *LDR* will not be as sensitive as the *Shuttle Infrared Telescope Facility* (*cf.*, Appendix C); even so, its very large size gives the *LDR* a tremendous advantage for problems requiring high spectral and angular resolution, such as studies of the atomic and molecular processes accompanying the formation of stars and planetary systems.

The technology for constructing this telescope is not yet available, but industry studies now underway are expected to advance technology sufficient to begin construction in the late 1980s; deployment in Earth orbit would follow in the early 1990s. By requiring performance only at infrared wavelengths longer than 0.03 millimeter, the required accuracy of the mirror surface and its supporting structure is greatly relaxed compared with that required for optical telescopes. In the long term, the experience gained in building and using the *LDR* should guide the design of very large optical and infrared telescopes that could be orbited in the 21st century. An artist's conception of the *LDR* being deployed from the *Space Shuttle* follows.

Courtesy of NASA

Very Long Baseline Array

Comprising ten 25-meter ground-based antennas widely distributed over United States territory, the *Very Long Baseline (VLB) Array* will fully exploit the powerful new technique of radio interferometry over long baselines. Still in the design stage, the *VLB Array* will likely include antennas in Hawaii and Puerto Rico, creating a maximum baseline of about 7500 kilometers – nearly 20 percent larger than Earth's radius. The figure below (courtesy of the National Radio Astronomy Observatory) shows the proposed locations of the individual antennas, each of which is an improved version of those comprising the 36-kilometer *Very Large Array* currently operating in New Mexico. Except for magnetic tape changes once a day, the antennas of the *VLB Array* will be largely unattended; instead, they will be controlled via telephone lines by a computer at an operations center at one of the antennas. At each site, hydrogen maser clocks will record timing markers to calibrate the received astronomical signals, after which the tapes will be shipped to the operations center for data analysis.

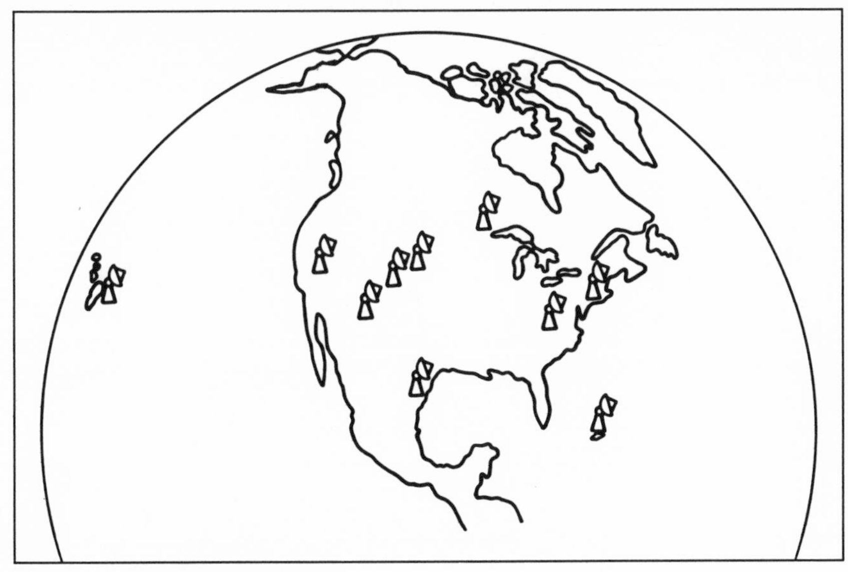

Courtesy of Dr. Kenneth Kellermann,
National Radio Astronomy Observatory

At its shortest operational wavelength (7 millimeters), the proposed *VLB Array* will achieve an angular resolution of roughly 0.0003 arc second, exceeding that of large ground-based optical telescopes by 3000 times and that of the *Space Telescope* by 200 times. (For comparison, 0.3 milliarc second equals the apparent size of a grain of sand on Earth as seen by an astronaut in Earth orbit.) Using multiple low-noise receivers, high-quality radio images will be obtainable at numerous wavelengths from 7 millimeters to 90 centimeters. The *VLB Array* will be able for the first time, among many other potential applications, to make continuous observations of the complex changes that occur within the small-scale structures surrounding the cores of quasars and active galaxies.

A long-term goal of radio astronomers is to place one or more antennas in deep elliptical orbits about the Earth, thereby extending the *VLB Array* over some 10^5 kilometers, and thus achieving an additional improvement of more than a factor of ten in angular resolution over any purely ground-based array. The required surface of such a space antenna is much less stringent than that of the *Large Deployable Reflector* (*cf.*, Appendix G) already under study; such a truly long-baseline interferometer would enable astronomers to probe angular scales of 10^{-5} arc second, where the rapid time variations in extragalactic radio sources imply the existence of meaningful structures.

Gamma-Ray Observatory

Now under construction for a projected launch by NASA's *Space Shuttle* in 1988, the *Gamma-Ray Observatory (GRO)* will carry telescopes having sensitivities and resolutions that greatly surpass those of previous detectors by at least an order of magnitude throughout the entire gamma-ray range. Far exceeding the capabilities of any previous gamma-ray mission, the *GRO* will be particularly adept at studying the properties of gamma-ray bursts. It will also accomplish, for the first time, routine study of gamma-ray emission lines in pulsars, supernovae, and the cores of active galaxies.

Shown below is an artist's conception of *GRO* shortly after launch by the *Space Shuttle.*

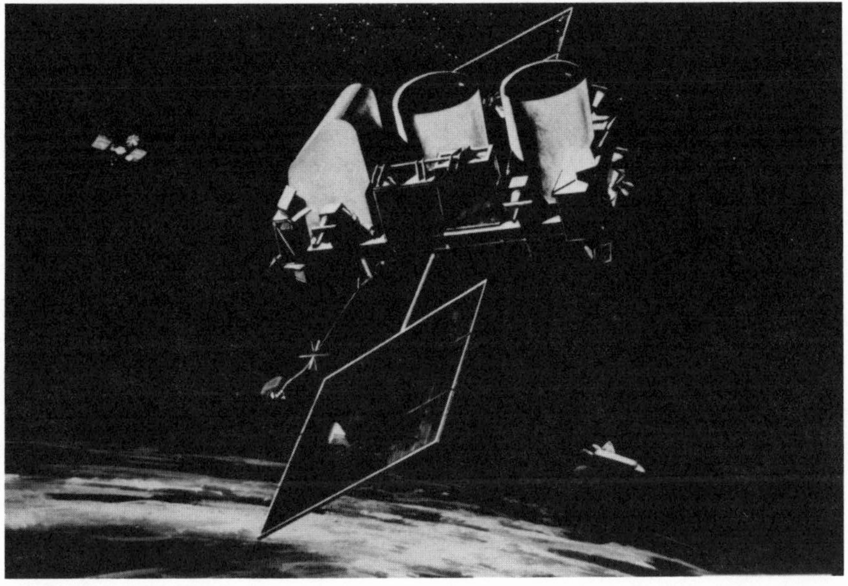

Courtesy of TRW, Inc.

Glossary

Absorption Feature (Line) A narrow wavelength interval in which radiation from an astronomical object has been absorbed by intervening material.

Abundance of the Elements The relative numbers of atoms of the various chemical elements, determined by study of their spectral lines in astronomical objects.

Accretion Disk A rotating disk of gas established when matter accreting onto an object is prevented from falling directly onto the object by its angular momentum, or tendency to continue rotating. The matter in the accretion disk slowly loses its angular momentum through friction, and spirals in to fall onto the object.

Active Galactic Nucleus A compact region in the center of a galaxy in which large amounts of energetic particles and radiation are produced, possibly by accretion onto a supermassive black hole. Examples include radio galaxies, Seyfert galaxies, and quasars.

Angstrom A unit of distance useful in measuring the wavelength of light, equal to 10^{-8} centimeters.

Angular Resolution The ability of an astronomical instrument to distinguish two objects close to one another in the sky. It is often stated in quantitative terms, such as seconds of arc.

Antimatter A form of matter complementary to ordinary matter, in which antiparticles (*e.g.*, antiprotons, positrons) have the opposite electric charge of the corresponding particles (*e.g.*, protons, electrons).

Arc Degree, Minute, Second Units for measuring angles designated by the word "arc" to avoid confusion with the corresponding units of time. A circle contains 360 arc degrees; a degree contains 60 arc minutes, and an arc minute contains 60 arc seconds.

Attenuation The loss of intensity of radiation due to absorption or scattering of radiation by the material through which it passes.

Big Bang A term used to describe the creation of the Universe, and its explosive expansion immediately following.

Binary System Two astronomical objects held in orbit around one another by their mutual gravitational attraction. Examples include binary stars and binary galaxies.

Black Hole An object in which all the mass is concentrated at a point. In its immediate vicinity the pull of gravity is so strong that light cannot escape–hence its name.

Boson One of the basic particles of nature, including the photon, the W, and the Z. Forces between other particles arise through the exchange of bosons.

CCD Charge-coupled device, an electronic detector used to detect astronomical images.

CID Charge-intensified device, an electronic detector used to detect astronomical images.

Cluster of Galaxies A system containing from a few dozen to a thousand or more galaxies bound together by their mutual gravitation.

Corona The very hot thin outer envelope of gas around a star.

Coronal Hole A region within the solar corona (*q.v.*) in which the density of gas is smaller than in the surrounding regions.

Cosmic Microwave Background Radiation An isotropic component of electromagnetic radiation thought to have originated at large distances in the Universe, at a time not long after the Big Bang. Its intensity has been best determined in the short radio wave or microwave part of the spectrum.

Cosmic Ray A particle striking the Earth and moving at close to the speed of light. Cosmic-ray particles include protons and helium nuclei, and are thought to be widespread throughout our Galaxy.

Cosmology The study of the structure and evolution of the entire Universe.

Dark Cloud A cloud of gas and dust in interstellar space which emits no light, and which appears dark because it attenuates the light from stars behind it.

Density The quantity of something in a unit of volume. Particle density is reckoned in particles per cubic centimeter, and mass density in grams per cubic centimeter.

Deuterium The chemical element "heavy hydrogen," whose nucleus, the deuteron, is made up of a proton and a neutron.

Doppler Shift The shift in wavelength upward (or downward) which results when a source of radiation moves away from (or toward) the observer. The Doppler shift depends upon the component of velocity along the line of sight.

DNA Deoxyribose nucleic acid, a long-chain organic molecule found in the nucleus of living cells; it carries the genetic information required for growth and reproduction.

Electromagnetic Radiation A form of energy that travels at the speed of light, 300,000 kilometers per second. The various types of electromagnetic radiation (x-ray, visible, radio, etc.) differ from one another only in their wavelengths.

Electron One of the basic particles of nature. It is stable and relatively light, and it carries a negative electrical charge.

Electroweak The force described by the unified theory of electromagnetic forces and weak nuclear forces.

Elliptical Galaxy A galaxy with an elliptical outline and a smooth distribution of light, normally containing little interstellar gas and dust.

Emission Feature (Line) A narrow wavelength interval in which radiation is produced by emission processes in a hot gas.

Emission Nebula A region in interstellar space containing hot gas which glows visibly.

Expansion of the Universe The observed phenomenon that almost all galaxies are receding from each other with velocities that are greater, the greater their distance. This expansion seemingly involves the entire Universe.

False Color An image-processing technique in which different images of an object taken at various different wavelengths are combined into a single image containing many colors.

Frequency The number of oscillations, or cycles, per second in a wave of radiation. One cycle per second is a unit of frequency called a Hertz.

Galactic Wind A stream of gas pouring out of a galaxy, believed to be responsible for removing interstellar gas from elliptical galaxies (*q.v.*).

Galaxy A giant system of stars far beyond the Milky Way. Most of the observed matter in the Universe is concentrated into galaxies, which are therefore considered as basic building blocks of the Universe.

Gamma Ray Electromagnetic radiation of extremely short wavelength and hence, extremely high frequency. Astronomical gamma rays are generated by only the very hottest and most energetic objects.

General Theory of Relativity A modern theory of the behavior of matter in the presence of gravitation. Conceived by Albert Einstein, it replaces Newton's concept of gravitational force by that of motion in curved space, the extent of curvature being determined by the distribution of matter in space.

Globular Cluster A group of up to a million stars held together by their mutual gravitation. Most globular clusters show evidence of containing only old stars, apparently formed soon after the Milky Way Galaxy (*q.v.*).

Grand Unified Theory A theory of matter which unifies the electroweak theory of electromagnetic and weak nuclear forces (*q.v.*) with the quantum chromodynamic theory of strong nuclear forces (*q.v.*). Still not fully tested by laboratory experiments, it makes predictions about what happened in the very early Universe.

Gravitational Instability A process in which small variations in the density of an otherwise uniform distribution of gas are amplified by the action of gravitation, which compresses the denser-than-average regions to higher density in a runaway manner. Gravitational instability was presumably involved in the formation of galaxies, and it is presently responsible for forming stars from interstellar gas and dust in our Galaxy.

Gravitational Radiation A novel type of radiation predicted by the general theory of relativity to be emitted whenever bodies move close to the speed of light under their mutual gravitational attraction. Not yet directly observed, it has been inferred indirectly from observations of a binary star system.

Heavy Element In this book, this phrase is used to denote chemical elements whose atoms are heavier than those of hydrogen and helium, the two lightest elements.

Heliosphere The region in space around the Sun in which the effects of the solar wind are felt strongly. It extends at least out to Saturn, and perhaps much further.

Helium The chemical element whose nucleus contains two protons and two neutrons. The simplest nucleus except for hydrogen, it is also the second most abundant element in the Universe (constituting about 10% of all atoms).

Hertz A unit of frequency, equal to one cycle per second.

Hidden Mass Matter in galaxies and in clusters of galaxies whose presence is inferred from its gravitational effects, but which is invisible.

Hubble's Law The relationship between the distance of a galaxy and its velocity of recession, determined by the Doppler shift of its spectrum. For velocities that are not too large, Hubble's Law states that the velocity is proportional to the distance. The constant of proportionality is the **Hubble constant,** whose reciprocal, the **Hubble time,** is an estimate for the age of the Universe.

Hydrogen The simplest chemical element, whose nucleus contains only a single proton; a rare isotope called deuterium (*q.v.*) also contains a neutron. About 90% of the atoms of ordinary matter in the Universe are hydrogen.

Image Processing System A computer and display terminal, together with operating programs, which enable astronomical images in digital form to be displayed and manipulated to obtain maximum scientific information.

Inflation The extremely rapid expansion of the Universe which grand unified theory predicts must have taken place 10^{-35} seconds after the origin of the Universe.

Infrared Electromagnetic radiation whose wavelength lies in the range between those of visible radiation at the short wavelength end, and radio waves at the long wavelength end.

Intercloud Medium A hot gas which pervades the large volumes of space between the cooler interstellar clouds in our Galaxy.

Interferometer A number of radio telescopes arranged so that the interference between the radio waves arriving at each telescope can be used to infer the direction of sources of radiation, and thus, to construct a radio image of a region of the sky.

Intergalactic Gas Gas in intergalactic space, outside of any galaxy. Absorption features in the spectra of quasars indicate that cool clouds of intergalactic gas exist; evidence for a hot intercloud component of intergalactic gas is controversial.

Interstellar Dust Small solid particles of matter in interstellar space whose presence is inferred from the attenuation of light over a broad range of wavelengths.

Interstellar Magnetic Field A weak magnetic field that permeates all of interstellar space.

Interstellar Molecule A chemical combination of atoms forming a molecule that is observed in interstellar space, usually by means of its radio emission.

Ionize To remove one or more electrons from an atom, leaving it with a net positive electrical charge.

Ionosphere A layer of the atmosphere above about 50 kilometers altitude in which solar radiation has removed an electron from some of the atoms.

Isotope A subspecies of a chemical element distinguished by the number of neutrons in the nucleus (the element itself is defined by the number of protons in the nucleus).

Isotropy The property of being independent of direction. For example, the x-ray background is observed to have nearly equal intensities in all directions, and is therefore isotropic.

Light-Year The distance light travels in one year, approximately 10 trillion kilometers.

Magnetic Storm An event in which the magnetic field at the surface of the Earth is observed to vary rapidly. It is caused by the impact of the solar wind on the Earth's magnetic field.

Maser A phenomenon in which microwaves (short-wavelength radio waves) are amplified when radiation stimulates molecules to emit more radiation in the same direction. To have an effective maser, the molecular states whose deexcitation gives the maser radiation must be continually reexcited by some outside energy source.

Milky Way The Galaxy we live in, so named because the stars in it can be seen overhead on a clear dark night as a milky band running across the sky.

Model In this book the term is used to connote a picture of a phenomenon constructed using the laws of physics. For example, although it is not possible to obtain a detailed image of the Cygnus-X1 x-ray source, a detailed model of it consistent with our observations comprises a stellar black hole in orbit around a normal star, and matter streaming from the normal star onto an accretion disk around the black hole.

Multiple Mirror Telescope A novel design for optical telescopes having several mirrors rather than one, all focused on the same object.

Neutrino A subatomic particle that has small or vanishing rest mass and no electrical charge. It can penetrate vast amounts of matter, as it interacts only very weakly with other particles.

Neutron A subatomic particle having about the same mass as a proton, but no electrical charge.

Neutron Star A compact star about the size of a city, in which the atomic nuclei are compressed so tightly as to be nearly touching. Under these conditions, most of them break down into neutrons, whence the name.

Nucleus The central core of an atom, around which electrons orbit, made up of protons and neutrons.

Photodensitometer A machine that scans over a photographic image and records it in digital form for later processing in a computer.

Photon A quantum of electromagnetic energy.

Pixel In an image that has been stored in digital form, a small area of the image, or picture element.

Planetary Nebula A shell of gas ejected by a star in the course of its evolution, and heated by the ultraviolet radiation of the star to glow brightly.

Plasma An ionized gas, that is, one in which the electrons have been removed from the atoms and are free to move about.

Positron The positively-charged antiparticle of the electron.

Precession The slow movement of the axis of rotation of a rapidly spinning object, such as a toy top or a rotating star.

Protostar A cloud of gas which is in the process of contracting to become a star.

Pseudocolor Color in a computer-produced image which is being used to represent intensity rather than true color.

Pulsar An astronomical object that appears to pulsate rapidly because it is spinning rapidly and its radiation is beamed. Most pulsars are neutron stars (*q.v.*).

Quantum Chromodynamics The theory that describes quarks (*q.v.*) and their interactions.

Quantum Electrodynamics The theory that describes electrically charged particles and their electromagnetic interactions.

Quantum Theory In general, a theory that incorporates the Heisenberg Uncertainty Principle governing all interactions. Thus, a quantum theory of gravitation would, if successful, extend Einstein's general theory of relativity to incorporate the Uncertainty Principle.

Quark A basic particle of which three combine to make a proton or a neutron.

Quasar Short for quasistellar radio source, an object whose large Doppler shift shows it to be distant in the Universe, but which has a pointlike image, like a star, rather than an extended one, like a galaxy. Quasars are thought to represent extremely energetic outbursts of active galactic nuclei.

Radioactivity A process in which atomic nuclei spontaneously disintegrate over time, emitting particles such as helium nuclei, electrons, positrons, gamma rays, and neutrinos.

Reconnection A process in which a magnetic field in space changes its structure, the magnetic lines of force reconnecting to each other in new ways.

Red Giant A cool star with a large diameter and high luminosity.

Relativistic As used in this book, moving with a velocity close to that of light.

Shock Wave A discontinuous pressure disturbance moving through matter at high speed.

Solar Activity The name given to a wide variety of phenomena on the surface of the Sun (such as the solar wind, sunspots, and flares) all of which appear to be governed by the solar magnetic field.

Solar Flare The rapid brightening of a small area of the solar surface, thought to be caused by the sudden conversion of magnetic energy into the energy of fast particles.

Solar Oscillation The repetitive up and down motion of the surface of the Sun, apparently caused by waves propagating within it.

Solar Prominence A loop of cool gas which is suspended, apparently by magnetic forces, above the surface of the Sun.

Solar Wind The expansive motion of the solar corona (*q.v.*) into space at velocities of about 500 kilometers per second. It never stops, but it is often intensified for limited times, in what are called "high-speed solar wind streams."

Spectrograph An instrument that disperses radiation into its component wavelengths, enabling one to study features such as absorption and emission lines (*q.v.*).

Spectrum A range of wavelengths, or alternatively, the distribution of intensity over a range of wavelengths.

Spiral Galaxy A galaxy which, like our own, has spiral arms in which interstellar gas and dust as well as young stars are concentrated.

Star Probe The name given by NASA to a proposed flight of a spacecraft very close to or even into our nearest star, the Sun.

Stellar Wind The expansion of the corona of a star, presumed to be analogous to the solar wind (*q.v.*).

Strong Nuclear Force The force that holds protons and neutrons together in an atomic nucleus. The energy released when the strong nuclear force draws protons and neutrons together is the source of energy for the Sun and stars.

Subatomic Particles Any of the particles comprising atoms (namely protons, neutrons, and electrons) or other particles that can be produced from them (such as neutrinos and positrons).

Supercluster of Galaxies A very large group of galaxies, up to hundreds of millions of light-years across, apparently held together by its own gravitation.

Super Grand Unified Theory A theory not yet constructed that would unify the gravitational force with the other three: electromagnetic force, weak nuclear force, and strong nuclear force.

Supernova The cataclysmic explosion of an entire star, giving rise to an outburst of radiation, a collapsed object such as a stellar black hole or neutron star, and a shock wave in interstellar space.

Supernova Remnant The cloud of glowing gas which is observed in space following a supernova explosion. The Crab Nebula, created by the supernova explosion of 1054 A.D., is the most famous one.

Thermonuclear Burning The reaction of atomic nuclei as a consequence of the high temperature at the center of a star, and the consequent ability of the nuclei involved to come close enough to react in spite of their electrical repulsion. Thermonuclear burning is the energy source for most stars.

Turbulence A state of motion of a medium such as water or air in which the flow pattern is so complex as to defy description except in a statistical manner.

Twenty-One Centimeter Line A spectral feature of atomic hydrogen at 21-centimeters wavelength, used to trace interstellar hydrogen gas in the Galaxy.

Ultraviolet Electromagnetic radiation having wavelengths shorter than those in the visible spectrum, but longer than those of x-rays.

Very Long Baseline Interferometer Two or more telescopes at great distances from one another whose observations of a single object in the sky can be combined in a computer to yield very precise positions and very detailed images.

Visible Spectrum The range of wavelengths in the electromagnetic spectrum, between 4000 and 7000 Angstroms, which is visible to the human eye.

Wavelength As pertains to radiation, the distance from the crest of a wave to that of the next succeeding wave.

Weak Nuclear Force A force between protons and neutrons that governs certain types of radioactivity. It is far weaker than the strong nuclear force.

White Dwarf A compact star about the size of the Earth, which is the remnant of the core of a star that has exhausted its nuclear fuel.

Window In this book, a range of the spectrum in which the attenuation by the Earth's atmosphere does not prevent astronomical observations (even though the attenuation may be severe at both shorter and longer wavelengths).

X Boson The particle that carries the force between other particles in a grand unified theory.

X Ray Electromagnetic radiation whose wavelength is shorter than that of ultraviolet radiation, but longer than that of gamma radiation. X rays are emitted only by very hot or energetic objects.

X-Ray Background An isotropic (*q.v.*) distribution of x-ray emission which is thought to originate at large distances in the universe, but whose source is unknown.

X-Ray Burster An object that exhibits sporadic bursts of x-ray emission, probably because of intermittent bursts of thermonuclear burning on the surface of a neutron star.

Suggestions for Further Reading

1. Radiation Visible and Invisible

Friedman, H., *The Amazing Universe,* 1975, National Geographic Society, Washington, D.C.

2. Interstellar Space

Bok, B. and Bok, P. F., *The Milky Way,* 1979, Harvard Univ. Press, Cambridge, Mass.

3. Sun and Stars

Eddy, J. A., *The New Sun,* 1979, NASA SP 402, U. S. Government Printing Office.

Noyes, R., *The Sun, Our Star,* 1983, Harvard Univ. Press, Cambridge, Mass.

Jastrow, R., *Red Giants and White Dwarfs,* 1979, Norton, New York.

4. Planets, Life, and Intelligence

Whipple, F., *Orbiting the Sun,* 1981, Harvard Univ. Press, Cambridge, Mass.

Billingham, J., ed., *Life in the Universe,* 1981, MIT Press, Cambridge, Mass.

5. Galaxies

Ferris, T., *Galaxies,* 1982, Sierra Club.

Mitton, S., *Exploring the Galaxies,* 1976, Scribner's, New York.

6. Cosmic Violence

Shipman, H. L., *Black Holes, Quasars, and the Universe,* 1980, Houghton Mifflin, Boston.

Silk, J., *The Big Bang,* 1980, Freeman, San Francisco.

7. Universe

Singh, J., *Great Ideas and Theories of Modern Cosmology,* 1970, Dover, New York.

Harrison, E. R., *Cosmology,* 1981, Cambridge Univ. Press.

8. Forces of Nature

Weinberg, S., *The First Three Minutes,* 1977, Basic Books, New York.

Trefil, J. S., *The Moment of Creation,* 1983, Scribner's, New York.

Calder, N., *The Key to the Universe,* 1977, Viking, New York.

Index

ıge numbers in italics refer to glossary definitions.